# Specification for piling and embedded retaining walls

- **Specification**
- **Contract documentation and measurement**
- **Guidance notes**

The Institution of Civil Engineers in collaboration with the Highways Agency, Ove Arup & Partners, and the Federation of Piling Specialists

THE INSTITUTION OF CIVIL ENGINEERS

ARUP

HIGHWAYS AGENCY

FPS

Thomas Telford

Brian Bell Associates
Upton House
4 Baldock St
Royston
Herts SG8 5AY

Published by Thomas Telford Publishing, Thomas Telford Services Ltd, 1 Heron Quay, London E14 4JD

First published 1996
Reprinted 1997, 2000, 2001, 2003

Distributors for Thomas Telford books are
*USA:* American Society of Civil Engineers, Publications Sales Department, 345 East 47th Street, New York, NY 10017–2398
*Japan:* Maruzen Co. Ltd, Book Department, 3–10 Nihonbashi 2-chome, Chuo-ku, Tokyo 103
*Australia:* DA Books and Journals, 648 Whitehorse Road, Mitcham 3132, Victoria

The Institution of Civil Engineers Specification is based on the Highways Agency Series 1600 Specification Piling and Embedded Retaining Walls which is Crown copyright and has been adapted with the permission of the Controller of Her Majesty's Stationery Office and the Highways Agency.

A catalogue record for this book is available from the British Library

**Classification**
*Availability:* Unrestricted
*Content:* Recommendations based on current practice
*Status:* Refereed
*User:* Practising civil and structural engineers and designers

ISBN: 0 7277 2566 1

Typeset by WestKey Ltd, Falmouth, Cornwall.

Printed and bound in Great Britain by The Cromwell Press, Trowbridge, Wiltshire.

# Foreword

The Institution of Civil Engineers *Specification for piling*, published in 1988, provided a new standard documentation for the different piling construction techniques. It was the wish of the Working Party which produced it that it should be quoted as a standard specification in contract documentation in the same way as the then Department of Transport *Specification for highway works*.

This new Specification includes many significant changes and it is believed that it can now be adopted as the *de facto* national standard for deep foundation works in the UK. Its origins started with a working party in the ICE whose task was to extend the 1988 specification to cover the increasingly important field of embedded retaining walls. In 1992, Ove Arup and Partners were commissioned by the now renamed Highways Agency to review completely the Agency's *Specification for piling and diaphragm walling*. Ove Arup and Partners asked the ICE for permission to use the draft developed by the working party on embedded walls and to use the existing ICE Specification as a starting point. In addition, they were asked by the Highways Agency to ensure that the specialist contractors, through their trade federation, the Federation of Piling Specialists, were consulted throughout the development. The result is a new Highways Agency document which provides a comprehensive specification for most load-bearing and embedded retaining wall techniques.

The Ground Board of the ICE considered that the Highways Agency documentation could be revised to leave intact the technical content of the specification but to remove references to specific Agency standards and documentation, and thereby to create a national standard specification. The Highways Agency without hindrance agreed to the ICE using their new documentation and this generous contribution is acknowledged by the inclusion of their logo on the ICE documentation. The ICE Specification presented herein is the end product of this collaboration. This is considered to be a significant step forward for the UK construction industry as it will mean that all deep foundation works can be undertaken to a common set of rules and expectations, with specialist contractors able to develop their own works procedures to a universally usable standard.

The first part of this volume comprises the *Specification for piling and embedded retaining walls*. It is anticipated that specifiers will refer only to this Specification and that it does not need to be reproduced for each contract. The document needs to contain only the special amendments to the Specification that the specifiers wish to make and the Particular Specification details called up by each of the relevant chapters. The second part of the volume contains guidance on particular issues (substantially based on the Contract Documentation and Measurement volume of the ICE *Specification for piling*, 1988), and the third part contains specific user notes for each chapter based on the Highways Agency *Notes for guidance*. Parts two and three would not form part of a contractual specification, but the advice contained in them would normally be reflected in contract documentation.

It is the aim of the Institution of Civil Engineers that this Specification should become the national *Specification for piling and embedded retaining walls*. It is acknowledged that there are many who would have a valuable contribution to make who have not been consulted in the evolution of this Specification. Anybody with comments, corrections or improvements which they would like to suggest for consideration for inclusion in the next edition of this Specification should write to the Chairman of the Ground Board at the Institution of Civil Engineers. To allow sufficient time for a body of experience with this Specification to develop, a seminar will be held at the end of 1998 to review its content and to nominate a working party to revise the Specification.

The co-operation between designers and specialist contractors and that of the Highways Agency in the development of this documentation has been exceptional, and is surely a model of how construction should be undertaken. As with any task of this nature, the amount of time and effort expended in the voluntary service given by the working parties and those providing constructive comment has been unstinting, and the contributions are most gratefully acknowledged.

**R H Wright**
**Chairman, ICE Ground Board**

# Acknowledgements

1988 Working Party for ICE *Specification for piling*
**S. Thorburn**, OBE, FEng, FICE, FIStructE, FASCE, MConsE; **W. G. K. Fleming** BSc, PhD, MICE; **J. Bickerdike** BSc, DIC, FICE; **A. Fawcett** BSc, PhD: **D. J. Palmer**, MA, FICE, MConsE; **J. May** BSc(Eng); **I. H. McFarlane** BSc(Eng); **D. Waite**, FIStructE, MICE; **J. Woodhouse**.

ICE Working Group for draft *Specification for embedded retaining walls*
**F. R. D. Chartres**, BSc, CGeol, MICE; **J. D. Findlay (Chairman)**, BSc, MSc, MICE; **E. T. Haws**, MA, FICE, FIPENZ; **D. E. Sherwood**, BSc, FICE; **V. M. Troughton**, BSc, MICE.

Highways Agency *Specification for piling and embedded retaining walls*, 1994
**D.I. Bush**, BSc, PhD, MICE; **T. J. P. Chapman**, BE, MSc, MIEI, MICE; **K. W. Cole**, BSc, MSc, FICE; **B. Simpson**, BA, MA, PhD, FICE; **V. M. Troughton**, BSc, MICE, **D. Twine**, BSc, MSc, MICE. This specification benefited from earlier contributions by **J. Mitchell** and **A. Turner**.

FPS Mirror Group for Highways Agency *Specification*
**A. Fawcett**, BSc, PhD; **J. Findlay**, BSc, MSc, MICE; **W. G. K. Fleming (Chairman)**, BSc, PhD, MICE; **L. Stansfield**, FICE, FIStructE.

ICE *Specification for piling and embedded retaining walls*, 1996
**T. J. P. Chapman**, BE, MSc, MIEI, MICE; **J. Findlay**, BSc, MSc, MICE; **J. A. Lord**, BA, MA, PhD, FICE, MConsE; **M. Alexander**.

# Contents

# Contract documentation and measurement 117

# Specification

# 1. General requirements for piling work

## 1.1. Definitions and Standards
### 1.1.1. Definitions

In this Specification the terms 'submitted', 'demonstrated', 'notified' and 'required' mean 'submitted to the Engineer', 'demonstrated to the Engineer', 'notified to the Engineer' and 'required by the Engineer' respectively.

**Allowable pile capacity**: a capacity which takes into account the pile's bearing capacity, the materials from which the pile is made, the required load factor, settlement, pile spacing, downdrag, the overall bearing capacity of the ground beneath the piles and other relevant factors. The allowable pile capacity indicates the ability of the pile to meet the specified loading requirements and is therefore required to be not less than the specified working load.

**Commencing surface**: is the level at which the piling equipment first enters the ground. This need not be the same as the piling platform level.

**Compression pile**: a pile which is designed to resist an axial force such as would cause it to penetrate further into the ground.

**Constant rate of penetration (CRP) test**: a test in which the pile is made to penetrate the soil at a constant controlled speed, while the force applied at the top of the pile to maintain the rate of penetration is continuously measured.

**Constant rate of uplift (CRU) test**: a test in which the pile is made to lift out of the soil at a constant controlled speed, while the force applied at the top of the pile to maintain the rate of extraction is continuously measured.

**Cut-off level**: is the level to which the pile concrete is trimmed in order to connect the pile to the structure.

**Design Verification Load (DVL)**: a load which will be substituted for the Specified Working Load for the purpose of a test and which may be applied to an isolated or singly loaded pile at the time of testing in the given conditions of the Site. The load will be particular to each Preliminary or other test pile and should take into account the maximum Specified Working Load for a pile of the same dimensions and materials, allowances for soil induced forces such as downdrag (which may act in reverse under temporary loading conditions), pile spacing and any other particular conditions of the test such as a variation of pile head casting level.

**Kentledge**: dead load used in a loading test

**Load**: a force applied to a pile under test or in service.

**Load factor**: the ratio between the pile's ultimate bearing capacity and maximum safe bearing capacity. The load factor relates only to the failure of the pile–soil interface. The allowable pile capacity is determined from the maximum safe bearing capacity making all necessary allowances.

**Maintained load test**: a loading test in which each increment of load is held constant either for a defined period of time or until the rate of movement (settlement or uplift) falls to a specified value.

**Pile settlement**: The axial movement at the top of the pile from the

*Specification for piling and embedded retaining walls.* Thomas Telford, London, 1996.

3

position before the commencement/start of loading. For piles loaded and unloaded through a number of cycles, settlement shall be the cumulative vertical movement.

**Preliminary pile**: a test pile installed before the commencement of the main piling works or specific part of the Works.

**Proof load**: a load applied to a selected working pile to confirm that it is suitable for the load at the settlement specified. A proof load should not normally exceed the Design Verification Load plus 50% of the Specified Working Load.

**Raking pile**: a pile installed at an inclination to the vertical.

**Reaction system**: the arrangement of the kentledge, piles, anchorages or spread foundations that provide a resistance against which the pile is load tested.

**Resistance**: the force developed by a pile in response to a load.

**Specified Working Load (SWL)**: the specified load on the head of the pile as stated in the relevant Particular Specification.

**Tension pile**: a pile which is designed to resist an axial force such as would cause it to be extracted from the ground.

**Test pile**: any pile to which a test load is, or is to be, applied.

**Ultimate bearing capacity**: the maximum resistance offered by the pile when the strength of the soil is fully mobilized. If the ultimate bearing capacity is to be derived from a test in which the displacement is not large enough to mobilize the strength of the soil fully, the ultimate bearing capacity may be derived from an extrapolation of the test results.

**Working pile**: one of the piles forming the foundation of a structure.

### 1.1.2. British Standards and other Codes of Practice

All materials and workmanship shall be in accordance with the appropriate British Standards, Codes of Practice and other specified standards current at the date of tender except where the requirements of these Standards or Codes of Practice are in conflict with this Specification in which case the requirements of this Specification shall take precedence.

### 1.2. Particular Specification

The following matters are, where appropriate, described in the Particular Specification:

(a) supervising officer
(b) location and description of the site
(c) nature of the works
(d) working area
(e) sequence of the works
(f) other works proceeding at the same time
(g) contract drawings
(h) office and other facilities for the Engineer
(i) submission of information (in addition to Table 1.1)
(j) responsibility for design
(k) performance criteria for the structure to be supported on the piles
(l) performance criteria for piles under test (see Table 1.3)
(m) permissible damage criteria for existing critical structures or services
(n) site datum and site grid
(o) restrictions on permissible working hours
(p) details of ground investigation reports
(q) additional ground investigation
(r) commencing surface
(s) other particular requirements

## 1.3. Progress Report

The Contractor shall submit to the Engineer on the first day of each week a progress report showing the current rate of progress and progress during the previous period on all important items of each section of the Works.

## 1.4. Pile layout, design and construction

*Option 1 — Contractor design*

The Contractor is required to design and construct piles having the qualities of materials and workmanship specified and which upon testing meet the requirements of the Specification for load–settlement behaviour. The Contractor's design shall comprise the calculation of individual pile lengths based on the ground conditions revealed by the site investigation to carry the Specified Working Loads within the specified limits for load–settlement behaviour.

The Contractor shall provide with his tender a schedule of sizes and lengths of the working piles including reinforcement and their corresponding allowable capacities to meet the requirements of the Specification.

*Option 2 — Engineer design*

The Contractor is required to construct piles of the type(s) and dimensions specified and having the qualities of materials and workmanship specified. Before the commencement of installation of the working piles the Contractor shall provide the Engineer with a schedule of pile sizes and lengths.

## 1.5. Materials

The sources of supply of materials shall not be changed until the Contractor has demonstrated that the materials from the new source can meet all the requirements of the Specification.

Materials failing to comply with the Specification shall be removed promptly from the site.

## 1.6. Safety
### 1.6.1. Standards

Safety precautions shall comply with all current legislation including the Health and Safety at Work Act 1974 or any subsequent re-enactment thereof and with BS 8004 and BS 8008.

### 1.6.2. Life-saving appliances

The Contractor shall provide and maintain on the Site sufficient proper and efficient life-saving appliances. The appliances must be conspicuous and available for use at all times.

Site operatives shall be conversant with the use of safety equipment and drills shall be carried out sufficiently frequently to ensure that all necessary procedures can be correctly observed.

### 1.6.3. Diving

Diving operations shall be carried out in accordance with the Diving Operations at Work Regulations (1981), Health and Safety Executive Publication 399, and any amendments or additions thereto.

Before any diving is undertaken the Contractor shall supply the Engineer with two copies of the code of signals to be employed, and shall have a copy of the Code prominently displayed adjacent to the diving control station on the craft or structure from which any diving operation takes place.

## 1.7. Ground conditions

No responsibility is accepted by the Engineer or Employer for any opinions or conclusions given in any factual or interpretative ground investigation reports. The Contractor shall report immediately to the Engineer any circumstance which indicates that in the Contractor's opinion the ground conditions differ from those

*Specification for piling and embedded retaining walls.* Thomas Telford, London, 1996.

5

Table 1.1. *Submission of information. (The following submissions shall be made to the Engineer at the time stated. Detailed requirements are listed under the clause number indicated in the relevant position.)*

| Section | Item | At Tender | Prior to commencing the Works | During the Works |
|---|---|---|---|---|
| 1 | Progress Report | | | 1.3 |
| | Pile layout, design and construction | 1.4.1 | | |
| | Piling method | 1.9 | | |
| | Piling programme | 1.10 | 1.10 | 1.10 |
| | Records | | | 1.11 |
| | Monitoring surveys | | 1.12.2 | |
| | Piling sequence | | 1.12.3 | |
| | Supervisor | | 1.13 | |
| | Quality Plan | | 1.13 | |
| 2 | Material test records | | | 2.3 |
| | Pile joints | | 2.4.2 | |
| | Means of maintaining concrete cover | | 2.6 | |
| | Prestressing records | | | 2.9.3 |
| | Grouting records | | | 2.9.12 |
| | Pile quality | | | 2.10 |
| | Handling, loading of piles | | 2.12 | |
| | Performance of driving equipment | | 2.13.3 | |
| | Length of piles | | | 2.13.4 |
| | Driving procedure | | | 2.13.5 |
| | Uplift and lateral displacement | | | 2.13.7 |
| 3 | Permanent casing | 3.6.2 | | |
| | Stability of pile bore | | 3.6.3 | |
| | Continuity of construction | | 3.6.6 | |
| | Means of maintaining concrete cover | | 3.7 | |
| | Pressure grouting method | | 3.12.2 | |
| | Uplift reference frame | | 3.12.4 | |
| | Pressure grouting records | | | 3.12.6 |
| 4 | Monitoring grout properties | | 4.8.2 | |
| | Monitoring system and records | | 4.10 | 4.10 |
| 5 | Performance of driving equipment | | 5.7.2 | |
| | Length of piles | | 5.7.3 | |
| | Driving procedure | | | 5.7.4 |
| | Uplift and lateral displacement | | | 5.7.6 |
| | Means of maintaining concrete cover | | 5.11 | |
| 6 | Ordering of piles | | 6.3 | |
| | Inspection and test certificates | | | 6.4.8 |
| | Welding procedures | | 6.6 | |
| | Manufacturing processes | | 6.7 | |
| | Weld tests | | | 6.8.2 |
| | Protection coats | | | 6.9.9 |
| | Performance of driving equipment | | 6.11.2 | |
| | Length of piles | | | 6.11.3 |
| | Driving procedure | | | 6.11.4 |
| | Uplift and lateral displacement | | | 6.11.6 |
| 7 | Origin of hardwood | | 7.3.1 | |
| | Inspection | | 7.4 | |
| | Treatment with preservative | | 7.5 | 7.5 |
| | Performance of driving equipment | | 7.9.2 | |
| | Length of piles | | | 7.9.3 |
| | Driving procedure | | | 7.9.4 |
| | Uplift and lateral displacement | | | 7.9.7 |
| 8 | Friction reduction method | 8.1 | | |
| | Pre-applied coating | | 8.3.1 | |
| | Pre-applied sleeving | | 8.4 | |
| | Formed-in-place low-friction surround | | 8.5 | |
| | Pre-installed sleeving | | 8.6 | |
| | Repair of damage | | | 8.7 |
| 9 | Specialist testing contractor | | 9.1.5 | |
| | Integrity testing report | | | 9.1.7 |
| | Anomolous results | | | 9.1.8 |
| | Calibration certificates | | | 9.2.2 |
| | Dynamic Testing results | | | 9.2.8 |
| 10 | Load application system | | | 10.1 |
| | Notice of start of construction | | | 10.3.1 |
| | Inclined piles for tension tests | | | 10.6.2 |
| | Use of working piles as reaction | | 10.6.3 | |
| | Adequate reaction | | | 10.6.5 |
| | Measurement of load | | | 10.8 |
| | Notice of test | | | 10.12 |
| | Results | | | 10.14.1 |
| | Recorded data | | | 10.14.2 |

*Specification for piling and embedded retaining walls.* Thomas Telford, London, 1996.

*Table 1.1. Continued*

| Section | Item | At Tender | Prior to commencing the Works | During the Works |
|---|---|---|---|---|
| 19 | Instrumentation supplier etc. | | 19.2 | |
| | Inclinometers | | | 19.4 |
| | Calibration and data checking | | | 19.8.2 |
| | Report | | | 19.8.4 |
| | Specialist instrumentation contractor | | | 19.8.5 |
| 20 | Types of cement | | 20.2.1 | |
| | Certificates of cement conformity | | | 20.2.1 |
| | Water tests | | | 20.4.2 |
| | Concrete workabilility | | 20.6.2 | |
| | Evidence of ASR compliance | | 20.6.2 | |
| | Detailed information on concrete mix | | 20.6.3 | |
| | Trial mixes | | 20.7.2 | |
| | Workability of each batch | | | 20.8.2 |
| | Concrete cube tests | | | 20.8.3 |
| 21 | Support fluid mix | 21.1 | | |
| | Use and compliance | | 21.3 | |

reported in or which could have been inferred from the ground investigation reports or preliminary pile results.

## 1.8. Tolerances
### 1.8.1. Setting out

Marker pins for the pile positions shall be set out and installed by the Contractor. Immediately prior to installation of the piles, the pile positions shall be checked by the Contractor.

### 1.8.2. Position

For a pile with a specified cut-off level at or above the commencing surface the maximum permitted deviation of the pile centre from the centre-point shown in the setting-out shall be 75 mm in any direction at commencing surface level. An additional tolerance for a pile with a specified cut-off level below commencing surface will be permitted in accordance with Clauses 1.8.3 and/or 1.8.4.

### 1.8.3. Verticality

The maximum permitted deviation of the finished pile from the vertical at any level is 1 in 75.

### 1.8.4. Rake

The maximum permitted deviation of any part of the finished pile from the specified rake is 1 in 25 for piles raking up to 1 in 6 and 1 in 15 for piles raking more than 1 in 6.

### 1.8.5. Forcible corrections to piles

Forcible corrections to concrete piles to overcome errors of position or alignment shall not be made. If forcible corrections are made to steel piles the Contractor shall demonstrate that the integrity, durability and performance of the piles have not been adversely affected.

## 1.9. Piling method

The Contractor shall submit with his tender all relevant details of the method of piling, the plant and monitoring equipment he plans to adopt. Alternative piling methods may be used provided it is demonstrated that they satisfy the requirements of the Specification.

## 1.10. Piling programme

The Contractor shall submit a provisional programme for the execution of the Works at the time of tender and a detailed programme prior to commencement of the Works. He shall inform the Engineer each day of the intended programme of piling for the following day and shall give 24 hours' notice of his intention to work outside normal hours and at weekends, where this is permitted.

*Specification for piling and embedded retaining walls.* Thomas Telford, London, 1996.

7

## 1.11. Records

The Contractor shall keep records as indicated by an asterisk in Table 1.2 for the installation of each pile and shall submit two signed copies of these records to the Engineer not later than noon of the next working day after the pile was installed. The signed records will form a record of the work.

Any unexpected driving or boring conditions shall be noted in the records.

*Table 1.2. Records to be kept (indicated by an asterisk)*

| Data | Pile type | | | | |
|---|---|---|---|---|---|
| | A | B | C | D | E |
| Contract | * | * | * | * | * |
| Pile reference number (location) | * | * | * | * | * |
| Pile type | * | * | * | * | * |
| Nominal cross-sectional dimensions or diameter | * | * | * | * | * |
| Nominal diameter of underream/base | - | - | - | * | - |
| Length of preformed pile | * | * | - | - | - |
| Standing groundwater level from direct observation or given site investigation data | - | - | * | * | * |
| Date and time of driving, redriving or boring from start to finish | * | * | * | * | * |
| Date of concreting | - | - | * | * | * |
| Ground level at pile position at commencement of installation of pile (commencing surface) | * | * | * | * | * |
| Working level on which piling base machine stands | * | * | * | * | * |
| Depth from ground level at pile position to pile toe | * | * | * | * | * |
| Toe level | * | * | * | * | * |
| Pile head level as constructed | * | * | * | * | * |
| Pile cut-off level | - | - | * | * | - |
| Length of temporary casing | - | - | * | * | - |
| Length of permanent casing | * | * | * | - | - |
| Type, weight, drop and mechanical condition of hammer and equivalent information for other equipment | * | * | * | - | - |
| Number and type of packing used and type and conditions of dolly used during driving of the pile | * | * | * | - | - |
| Set of pile or pile tube in millimetres per 10 blows or number of blows per 25 mm of penetration | * | * | * | - | - |
| Temporary compression of ground and pile | | | | | |
| Driving resistance taken at 0.25 m intervals | * | * | * | - | - |
| Soil samples taken and in-situ tests carried out during pile formation or adjacent to pile position | * | * | * | * | * |
| Length and details of and cover to reinforcement | - | - | * | * | * |
| Concrete mix | - | - | * | * | * |
| Volume of concrete supplied to pile where this is practical | - | - | * | * | * |
| All information regarding obstructions delays and other interruptions to the sequence of work | * | * | * | * | * |
| Pile forming equipment including rig no. | * | * | * | * | * |
| Description of ground excavated | - | - | - | * | * |
| Depth from commencing surface to changes in strata and to standing ground water and any fluctuations | - | - | - | * | * |
| Depth to average levels of concrete surface before and after withdrawing temporary lining | - | - | - | * | - |
| Type, torque, assessed efficiency of motor | - | - | - | - | * |
| Depth to concrete surface after every concrete load | - | - | * | * | - |
| For raking piles, angle of rake | - | - | - | * | - |
| Support fluid tests | * | - | - | - | - |
| Monitoring information referred to in Section 4.8 | - | - | - | * | - |
| Level of top of reinforcement cage, as constructed | - | - | * | * | * |

**Notes**

1.  Pile types:
    A   Driven precast concrete and steel piles
    B   Driven segmental concrete piles
    C   Driven cast-in-place concrete piles
    D   Bored cast-in-place concrete piles
    E   Continuous flight auger concrete or grout piles

2.  All levels shall be relative to the Datum specified in the Particular Specification

3.  All times shall be given in 24 hour format

*Specification for piling and embedded retaining walls.* Thomas Telford, London, 1996.

## 1.12. Nuisance and damage

### 1.12.1. Noise and disturbance

The Contractor shall carry out the work in such a manner and at such times as to minimize noise, vibration and other disturbance in order to comply with current environmental legislation.

Particular restrictions on permissible working hours are stated in the Particular Specification.

### 1.12.2. Damage to adjacent structures

Permissible damage criteria for adjacent structures or services are given in the Particular Specification. If in the opinion of the Contractor damage may be caused to other structures or services by his execution of the Works he shall immediately notify the Engineer. The Contractor shall submit his plans for making surveys and monitoring movements or vibration before the commencement of the Works.

### 1.12.3. Damage to piles

The Contractor shall ensure that during the course of the work, displacement or damage which would impair either performance or durability does not occur to completed piles.

The Contractor shall submit to the Engineer his planned sequence and timing for driving or boring piles, having regard to the avoidance of damage to adjacent piles.

### 1.12.4. Temporary support

The Contractor shall ensure that where required, any permanently free-standing piles are temporarily braced or stayed immediately after driving to prevent loosening of the piles in the ground and to ensure that no damage resulting from oscillation, vibration or movement can occur.

## 1.13. Supervision and control of the Works

The Contractor shall keep upon the Works a competent site supervisor to be in charge of pile construction and installation.

The site supervisor must be experienced in the type of pile construction necessitated by the Contract. A curriculum vitae of the supervisor shall be submitted prior to commencement. The whole time of the site supervisor shall be devoted to the piling works. The site supervisor shall not be removed from the Works without the Engineer being notified in advance with at least one week's notice.

The Contractor shall submit one week prior to commencement of piling works his Quality Plan for the Works. Subsequent revisions, amendments or additions shall be submitted prior to their implementation. Quality Assurance and Quality Control documentation shall be made available on request.

*Table 1.3. Performance criteria for piles*

| Pile Ref. Nos | Permitted type(s) — Specification Section No. | Specified working load | Pile designation | Design verification Load DVL (where applicable) | Load factor | Permitted settlement at DVL, mm | Permitted settlement at DVL + $\frac{1}{2}$SWL, mm | Minimum pile length from cut-off to toe | Minimum pile diameter or dimensions of cross-Section |
|---|---|---|---|---|---|---|---|---|---|
| | | | | (kN) | | (kN) | | | |
| | | | | | | Maximum settlement | Maximum settlement | | |

*Specification for piling and embedded retaining walls.* Thomas Telford, London, 1996.

9

# 2. Precast reinforced and prestressed concrete piles and precast reinforced concrete segmental piles

**2.1. General**

All materials and work shall be in accordance with Sections 1, 2 and 20 of this Specification, except where there may be conflict of requirements, in which case this Section shall take precedence.

**2.2. Particular Specification**

The following matters are, where appropriate, described in the Particular Specification:

- (*a*) specified working loads
- (*b*) additional performance criteria for piles under test (also see Table 1.3)
- (*c*) type of cement
- (*d*) types and sizes of aggregate
- (*e*) grades of concrete
- (*f*) designed or prescribed mixes and maximum free water to cement ratio
- (*g*) method of testing concrete workability
- (*h*) grades and types of and cover to reinforcement
- (*i*) types of prestressing tendon
- (*j*) grout
- (*k*) marking of piles
- (*l*) penetration or depth or toe level
- (*m*) driving resistance or dynamic evaluation or set
- (*n*) trial drives
- (*o*) preliminary piles
- (*p*) uplift/lateral displacement trials
- (*q*) pile shoes (where required)
- (*r*) preboring or jetting or other means of easing pile driveability
- (*s*) detailed requirements for driving records (including requirements for measurement of temporary compressions and redrives)
- (*t*) disposal of cut-off heads of piles
- (*u*) other particular requirements.

**2.3. Ordering of piles**

The Contractor shall ensure that the piles are available in time for incorporation in the Works. All piles and production facilities shall be made available for inspection at any time. Piles shall be examined at the time of delivery and any faulty units replaced. The records of testing of the concrete and steel used in the piles shall be submitted.

**2.4. Materials and components**
**2.4.1. Steel and iron components**

In the manufacture of precast concrete piles and jointed precast concrete segmental piles, fabricated steel components shall comply with BS 4360 Grades 43A or 50A or BS EN10 025 Grades Fe 430A or Fe 510A, cast steel components with BS 3100, grade A (A1, A2 or A3) and ductile iron components with BS 2789 Grades 420/12, 400/18 or 350/22.

### 2.4.2. Pile joints

The joints shall be close-fitting face to face and the locking method shall be such as to hold the faces in intimate contact. Details of the design, manufacture and tests of the jointing system shall be submitted prior to the commencement of the Contract.

A jointed pile shall be capable of withstanding the same driving stresses as a single unjointed pile of the same cross-sectional dimensions and materials.

The welding of a joint to main reinforcement in lieu of a lapped connection with projecting bars affixed to the joint shall not be permitted.

Each pile joint shall be square to the axis of the pile within a tolerance of 1 in 150. The centroid of the pile joint shall lie within 5 mm of the true axis of the pile element.

### 2.4.3. Pile toes

Pile toes shall be constructed so as to ensure that damage is not caused to the pile during installation.

### 2.4.4. Pile head reinforcement

Pile heads shall be so reinforced or banded as to prevent bursting of the pile under driving.

## 2.5. Tolerances in pile dimensions

The cross-sectional dimensions of the pile shall be not less than those specified and shall not exceed them by more than 6 mm. Each face of a pile shall not deviate by more than 6 mm from any straight line 3 m long joining two points on that face, nor shall the centre of area of the pile at any cross-section along its length deviate by more than $1/500$ of the pile length from a line joining the centres of area at the ends of the pile. Where a pile is less than 3 m long the permitted deviation from straightness shall be reduced below 6 mm on a *pro rata* basis in accordance with actual length.

The head of a pile element or the end of the pile upon which the hammer acts shall be square to the pile axis within a tolerance of 1 in 150.

## 2.6. Reinforcement
### 2.6.1. Precast reinforced and prestressed concrete piles

The main longitudinal reinforcing bars in piles not exceeding 12 m in length shall be in one continuous length. In piles more than 12 m long, joints will be permitted in main longitudinal bars so that the number of joints is minimized. Joints in reinforcement shall be such that the full strength of the bar is effective across the joint, following the guidance of BS 8110.

Lap or splice joints shall be provided with sufficient link bars to resist eccentric forces. Reinforcement shall be incorporated for lifting and handling purposes.

Spacers shall be designed and manufactured using durable materials which shall not lead to corrosion of the reinforcement or spalling of the concrete cover. Details of the means by which the Contractor plans to ensure the correct cover to and position of the reinforcement shall be submitted.

If a pile is constructed with a shaped point or shoe, then the end of the pile shall be symmetrical about the longitudinal axis of the pile. Holes for handling or pitching, where provided in the pile, shall be lined with steel tubes; alternatively, inserts may be cast in.

### 2.6.2. Precast reinforced concrete segmental piles

The main longitudinal reinforcing bars shall be in one continuous length. Sufficient reinforcement shall be incorporated for lifting and handling purposes.

Spacers shall be designed and manufactured using durable materials which shall not lead to corrosion of the reinforcement or spalling of the concrete cover. Details of the means by which the

Contractor plans to ensure the correct cover to and position of the reinforcement shall be submitted.

## 2.7. Formwork

Formwork shall be robust, clean and so constructed as to prevent loss of grout or aggregate from the wet concrete and ensure the production of uniform pile sections, free from defects. The piles shall be removed from the formwork in such a manner that significant damage to the pile is prevented.

## 2.8. Concrete
### 2.8.1. Placing and compacting concrete

The method of placing and compacting the concrete shall be such as to ensure that the concrete in its final position is dense and homogeneous.

### 2.8.2. Protecting and curing concrete

Immediately after compaction, concrete shall be adequately protected from the harmful effects of the weather, including wind, rain, rapid temperature changes and freezing. Curing shall be carried out in accordance with BS 8110.

Piles shall not be removed from formwork until a sufficient pile concrete strength has been achieved to allow the pile to be handled without significant damage.

When accelerated curing is used the curing procedure shall ensure no deleterious effects to the pile. Four hours must elapse from the completion of placing concrete before the temperature is raised. The rise in temperature of the concrete within any period of 30 minutes shall not exceed 10°C and the maximum temperature attained shall not exceed 70°C. The rate of subsequent cooling shall not exceed the rate of heating.

## 2.9. Prestressing
### 2.9.1. General

The Contractor shall ensure that the Engineer is given adequate notice and every facility for inspecting the manufacturing process.

Prestressing operations shall be carried out only under the direction of an experienced and competent supervisor. All personnel operating the stressing equipment shall have been trained in its use.

The extensions and total forces, including allowance for losses, shall be calculated before stressing commences.

Stressing of tendons and transfer of prestress shall be carried out at a gradual and steady rate. The force in the tendons shall be obtained from readings on a recently calibrated load cell or pressure gauge incorporated in the equipment. The extension of the tendons under the total forces shall be within 5% of the calculated extension.

### 2.9.2. Concrete strength

The Contractor shall cast sufficient cubes, cured in the same manner as the piles, to be able to demonstrate by testing two cubes at a time, with appropriate intervals between pairs of cubes, that the necessary in situ strength of the concrete has been reached.

Concrete in the piles shall not be stressed until two test cubes attain the necessary strength.

### 2.9.3. Records

The Contractor shall keep detailed records of times of tensioning, measured extensions, pressure gauge readings or load cell readings and the amount of pull-in at each anchorage. The Contractor shall also check the cover to reinforcement of each pile and shall ensure it is at least that specified. Copies of these records shall be submitted within such reasonable time from completion of each tensioning operation as may be required, and in any case not later than noon on the following working day.

*Specification for piling and embedded retaining walls.* Thomas Telford, London, 1996.

### 2.9.4. Post-tensioned piles

Apart from the requirement for ducts and vents in post-tensioned piles to be grouted after the transfer of prestress, the requirements for pre-stressed piles apply. Where specified in the Particular Specification, facilities shall be provided for retensioning piles after driving.

### 2.9.5. Grouting procedure

Grout shall be mixed for a minimum of 2 minutes and until a uniform consistency is obtained.

Ducts shall not be grouted when the air temperature in the shade is lower than 3°C.

Before grouting is started all ducts shall be thoroughly cleaned by means of compressed air which shall be free of oils.

Grout shall be injected near the lowest point in the duct in one continuous operation and allowed to flow from the outlet until the consistency is equivalent to that of the grout being injected.

Vents in ducts shall be provided in accordance with Clause 8.9.2 of BS 8110.

### 2.9.6. Grout

Grout for prestressing shall have the following properties.

(a) The design and workability of grout to be used in the formation of piles shall produce a mix which is suitable for pumping.

(b) Grout shall consist only of Portland cement, water and admixtures. Admixtures are only to be used in accordance with the Manufacturer's instructions and shall comply with Section 21.

(c) Grout shall have a water/cement ratio as low as possible consistent with the necessary workability, and the water cement ratio shall not exceed 0.40.

(d) Grout shall not bleed in excess of 2% after 3 hours, or a maximum of 4%, when measured at 18°C in a covered glass cylinder approximately 100 mm in diameter with a height of grout of approximately 100 mm, and the water shall be reabsorbed after 24 hours.

### 2.9.7. Testing Works grout

Cube strength testing shall be carried out in accordance with BS 1881. A sample shall consist of a set of six 100 mm cubes. Three cubes shall be tested at seven days and the remaining three at 28 days after casting.

### 2.9.8. Batching grout

The weighing and water-dispensing mechanisms shall be maintained at all times to within the limits of accuracy described in BS 1305.

The weights of each component of the mix shall be within 2% of the respective weights per batch.

### 2.9.9. Mixing grout

Cement grouts shall be mechanically mixed thoroughly to produce a homogeneous mix.

The grout shall be mixed on site until a homogeneous grout is obtained and shall be kept in continuous movement until it is used. It shall be used within 30 minutes from the start of mixing. Following mixing, the grout shall be passed through a 5 mm aperture sieve and shall be remixed if lumps are retained.

### 2.9.10. Transporting grout

Grout shall be transported from the mixer to the position of the pile in such a manner that segregation of the mix does not occur.

### 2.9.11. Placing grout in cold weather

Grout shall have a minimum temperature of 5°C when placed. No frozen material or material containing ice shall be used for making grout. All plant and equipment used in the transporting and placing of grout shall be free of ice that could enter the grout.

## 2.9.12. Records

The Contractor shall keep records of grouting, including the date, the proportions of the grout and any admixtures used, the pressure, details of interruption and topping up required. Copies of these records shall be submitted within such reasonable time after completion of each grouting operation as may be required, and in any case not later than noon on the following working day.

## 2.10. Pile quality

A certificate of quality from the pile manufacturer shall be provided stating that the requirements of this Specification have been fulfilled during manufacture.

## 2.11. Marking of piles

Each pile shall be marked in such a manner that it can be identified with the records of manufacture which shall state the date of casting, the cement type, concrete grade, element length, the prestressing force where appropriate and any other relevant data. On delivery the piles shall be accompanied by records of manufacture. Lifting positions shall be marked on each pile in accordance with the requirements of design.

## 2.12. Handling, transportation and storage of piles

The method and sequence of lifting, handling, transporting and storing piles shall be such as to avoid shock loading and to ensure that the piles are not damaged. Only designed lifting and support points shall be used. During transport and storage, piles shall be appropriately supported under the marked lifting points or fully supported along their length.

All piles within a stack shall be in groups of the same length. Packing of uniform thickness shall be provided between piles at the lifting points.

Concrete shall at no time be subjected to loading, including its own weight, which will induce a compressive stress in it exceeding 0.33 of its strength at the time of loading or of the specified strength, whichever is the lesser. For this purpose the assessment of the strength of the concrete and of the stresses produced by the loads shall be submitted.

A pile shall be rejected when the width of any transverse crack exceeds 0.3 mm. The measurement shall be made with the pile in its working attitude.

## 2.13. Driving piles
## 2.13.1. Strength of piles

Piles shall not be driven until the concrete cubes have achieved the specified characteristic strength and the pile has attained the strength to resist driving stresses, whichever is the greater.

## 2.13.2. Leaders and trestles

At all stages during driving and until incorporation into the structure the pile shall be adequately supported and restrained by means of leaders, trestles, temporary supports or other guide arrangements to maintain position and alignment and to prevent significant damage to the pile

## 2.13.3. Performance of driving equipment

The efficiency and energy of the driving equipment including when followers are used shall be notified. Where required in the Particular Specification, dynamic evaluation and analysis shall be provided.

Where a drop hammer is used, the mass of the hammer shall be at least half that of the pile. For other types of hammer the energy delivered to the pile per blow shall be at least equivalent to that of a drop hammer of the stated mass. Drop hammers shall not be used from floating craft in such a manner as to cause instability of the craft or significant damage to the pile.

### 2.13.4. Length of piles

The length of pile to be driven in any position shall comply with the Contractor's schedule unless otherwise specified in the Particular Specification. During the execution of the Works any amendments to the scheduled lengths shall be reported to the Engineer giving adequate notice prior to installation of the piles concerned.

### 2.13.5. Driving procedures and redrive checks

The driving procedure shall be such as to avoid damage to the piles.

The Engineer shall be given 24 hours' notice of commencement of the driving of the first pile.

The Contractor shall give adequate notice and provide all necessary facilities to enable the Engineer to check driving resistances, sets and temporary compressions.

The driving of each pile shall be continuous until the depth or set as required by the design has been reached. In the event of unavoidable interruption to driving, the pile may be redriven provided it can subsequently be driven to the designed depth and/or resistance or set without damage. A follower shall not be used unless the set is revised where applicable in order to take into account reduction in the effectiveness of the hammer blow.

Driving records shall be made for every pile. This record shall contain the weight and fall of the hammer or ram and the number of blows for each 0.25 m of penetration, unless otherwise specified in the Particular Specification.

The Contractor shall inform the Engineer without delay if an unexpected change in driving characteristics is noted.

### 2.13.6. Set

The set and temporary compression shall be measured and recorded for each pile at the completion of driving unless otherwise stated in the Particular Specification.

When a set is being measured, the following requirements shall be met.

(a) The exposed part of the pile shall be in good condition without damage or distortion.
(b) The helmet, dolly and any packing shall be in sound condition.
(c) The hammer blow shall be in line with the pile axis and the impact surfaces shall be flat and at right angles to the pile and hammer axis.
(d) The hammer shall be in good condition, delivering adequate energy per blow, and operating correctly.
(e) The temporary compression of the pile shall be recorded if required in the Particular Specification.

The set shall be recorded either as the penetration in millimetres per 10 blows or as the number of blows required to produce a penetration of 25 mm.

### 2.13.7. Driving sequence and risen piles

Piles shall be driven in a sequence to minimise any detrimental effects of heave and lateral displacement of the ground. The sequence and method of piling including preboring shall limit uplift and lateral movement so that the final position of each pile is within the specified tolerances. At all times the deflections of each pile from its axis as driven shall not be such as to cause damage or impair durability of the piles or any structures or services.

The maximum permitted uplift of each pile due to any one pile driven within a pile centre to centre radius of eight pile diameters is 3 mm unless it can be demonstrated by static load testing that uplift exceeding this amount does not affect the ability of the pile to meet the requirements of this Specification.

Even if during the installation of preliminary piles uplift is shown to be within the permitted maximum and the preliminary piles tested meet the requirements of the Specification checks of uplift on working piles shall be made by the Contractor at least once a week throughout the period of the piling work and the results reported to the Engineer.

If preliminary piles are not installed the Contractor shall commence installation of working piles taking measures to reduce or eliminate uplift until it can be established by site measurements that such measures are no longer necessary. Thereafter checks on uplift shall be made by the Contractor at least once a week and the results reported to the Engineer.

If a static load test shows that a pile which uplifted more than the maximum permitted amount due to any pile driven within a radius of eight pile diameters does not comply with the requirements of the Specification all such piles that may have been uplifted shall be redriven.

If records and measurements show that piles have been laterally displaced so as to be outside the permitted tolerance, damaged or are of impaired durability the measures the Contractor plans to adopt to enable the piles to comply with the Specification shall be provided to the Engineer.

Laterally displaced piles shall not be corrected by forcible correction at the heads.

### 2.13.8. Preboring

If preboring is specified, the diameter and depth of prebore shall be as stipulated in the Particular Specification.

Other means to ease pile drivability may be used provided the completed piles meet the requirements of the Specification.

## 2.14. Repair and lengthening of piles
### 2.14.1. Repair of damaged pile heads

If it is necessary to repair the head of a pile before it has been driven to its final level, the Contractor shall carry out such repair in a way which allows the pile-driving to be completed without further damage. If the driving of a pile has been completed but the level of sound concrete of the pile is below the required cut-off level, the pile shall be made good to the cut-off level, or the pile cap or substructure may be locally deepened, so that the completed foundation will safely withstand the specified working load.

### 2.14.2. Lengthening of precast reinforced and prestressed concrete piles

Any provision for lengthening piles incorporated at the time of manufacture shall be designed by the Contractor to resist all stresses to which it may be subjected.

If no provision for lengthening piles was incorporated at the time of manufacture, any method for lengthening shall be such that the extended pile including any joints is capable of taking safely the stresses during driving and under load.

### 2.14.3. Lengthening of precast reinforced concrete segmental piles

Where piles are driven to depths exceeding those expected, leaving insufficient projection for penetration into the following works, the piles shall be extended or replaced so that the completed piles are capable of meeting the requirements of the Specification.

### 2.14.4. Driving repaired or lengthened piles

Repaired or lengthened piles shall not be driven until cubes for the added concrete have reached the specified characteristic strength of the concrete of the pile.

## 2.15. Cutting off pile heads

When the driving of a pile has satisfied the Specification requirements the concrete of the head of the pile shall be cut off to the level specified. Reinforcing bars projecting above this level shall be as specified.

Care shall be taken to avoid shattering or otherwise damaging the rest of the pile. Any cracked or defective concrete shall be cut away and the pile repaired to provide a full and sound section to cut-off level.

# 3.  Bored cast-in-place piles

**3.1.  General**

All materials and work shall be in accordance with Sections 1, 3 and 20 of this Specification, except where there may be conflict of requirements, in which case this Section shall take precedence.

**3.2.  Particular Specification**

The following matters are, where appropriate, described in the Particular Specification:

- (*a*)  specified working loads
- (*b*)  additional performance criteria for piles under test (also see Table 1.3)
- (*c*)  types of cement
- (*d*)  cement replacement materials
- (*e*)  types and sizes of aggregate
- (*f*)  grades of concrete
- (*g*)  designed or prescribed mixes and maximum free water to cement ratio
- (*h*)  method of testing concrete workability
- (*i*)  grades and types of and cover to reinforcement
- (*j*)  permanent casing
- (*k*)  support fluid
- (*l*)  pile dimensions
- (*m*)  pressure grouting
- (*n*)  preliminary piles
- (*o*)  trial bores
- (*p*)  whether pile inspection by manned descent is required
- (*q*)  disposal of excavated material
- (*r*)  other particular requirements.

**3.3.  Support fluid**

Where support fluid is used to maintain the stability of the excavation it shall be in accordance with Section 21.

**3.4.  Setting out**

The Contractor shall check the casing position for each pile during and immediately after placing of the casing. Any checks by the Engineer shall not relieve the Contractor of his responsibility.

**3.5.  Diameter of piles**

The diameter of a pile shall be not less than the specified diameter.
    The auger width shall be checked as necessary and recorded for each pile to ensure the specified diameter is achieved. A tolerance of 5% on the auger width is permissible.

**3.6.  Boring**
**3.6.1.  Boring near recently cast piles**

Piles shall be bored in an order and in such a manner that no significant damage is sustained by previously formed piles.

**3.6.2.  Casings**

Temporary casings shall be of quality of material, length and thickness adequate for the purpose of preventing water and unstable soil from entering the pile excavations. A short length of temporary casing shall be provided for all piles to provide an upstand of at least 1 m above surrounding ground level for safety and to prevent contamination of the concrete in the bore.

Temporary casings shall maintain the excavations to their full dimensions and ensure that piles are completed to their full cross-sectional dimensions.

The use of a vibrator to insert and withdraw temporary casings is subject to compliance with Clause 1.12 and to the method not causing disturbance of the ground which would adversely affect the construction or the capacity of piles.

Temporary casings shall be free from significant distortion. They shall be of uniform cross-section throughout each continuous length. During concreting they shall be free from internal projections and encrusted concrete which might adversely affect the proper formation of piles.

Where piles are bored under water or support fluid in an unlined state, the insertion of a full-length loosely fitting casing to the bottom of the bore prior to placing concrete will not be permitted.

Permanent casings shall be as specified in the Particular Specification.

Where the use of a permanent casing is specified, the Contractor shall submit details of the type of casing and the method of installation at the time of tender.

### 3.6.3. Stability of pile bore

Where boring takes place through unstable water-bearing strata, the process of excavation and the support fluid and depth of temporary casing where employed shall be such that soil from outside the area of the pile is not drawn into the pile section and cavities are not created outside the temporary casing as it is advanced.

Where a support fluid is used for maintaining the stability of a bore, an adequate temporary casing shall be used in conjunction with the support fluid so as to ensure stability of the strata near ground level until concrete has been placed. During construction the level of drilling fluid in the pile excavation shall be maintained within the cased or stable bore so that it is not less than 2 m above the level of external standing groundwater at all times, unless stated otherwise in the Particular Specification.

In the event of a loss of support fluid from a pile excavation, the Contractor shall notify the Engineer of his intended action before continuing the work.

### 3.6.4. Spillage and disposal of support fluid

All reasonable steps shall be taken to prevent the spillage of support fluid on the Site in areas outside the immediate vicinity of boring. Discarded fluid shall be removed from the Site without undue delay. Any disposal of fluid shall comply with the requirements of current legislation and all relevant authorities.

### 3.6.5. Pumping from pile bores

Pumping from pile bores shall not be permitted unless the bore has been sealed against further water entry by casing or unless the soil is stable and will allow pumping to take place without ground disturbance below or around the pile.

### 3.6.6. Continuity of construction

The pile shall be bored and the concrete shall be placed without such delay as would lead to impairment of the performance of the pile.

The time period during which each pile is excavated and the concrete is placed shall not exceed 12 hours. The time period shall start when excavation below the temporary lining tubes commences (except for casings such as used in secant piling which leave a significant annulus between the casing and the soil). Where the construction sequence is such that the time period of 12 hours will be exceeded even if no delays are taken into account, a realistic

*Specification for piling and embedded retaining walls.* Thomas Telford, London, 1996.

19

time period during which the pile is excavated and concrete placed shall be stated in the Contractor's method statement. The Contractor shall advise on the likely effect of this extension of the pile construction period on the performance and capacity of the pile.

### 3.6.7. Enlarged pile bases

A mechanically formed enlarged base shall be no smaller than the dimensions specified and shall be concentric with the pile shaft to within a tolerance of 10% of the shaft diameter. The sloping surface of the frustum forming the enlargement shall make an angle to the axis of the pile of not more than 35°.

At the specified diameter of the underream at the perimeter of the base there shall be a minimum height of 150 mm.

### 3.6.8. Cleanliness of pile bases

On completion of boring loose, disturbed or softened soil shall be removed from the bore using appropriate methods, which shall be designed to clean while at the same time minimizing ground disturbance below the pile bases. Where used, support fluid shall be maintained at such levels throughout and following the cleaning operation that stability of the bore is preserved. At all times when the pile head is unattended, the bore shall be clearly marked and fenced off so as not to cause a safety hazard.

### 3.6.9. Inspection

Each pile bore which does not contain standing water or support fluid shall be inspected from the ground surface prior to concrete being placed in it to ensure the base is adequately clean and that the shaft is stable and within the specified tolerances. Adequate means of lighting, measuring tapes and a means of measuring verticality shall be used. For piles of 750 mm diameter or larger and where manned inspection of the pile base is specified in the Particular Specification, equipment shall be provided by the Contractor to enable his representatives to descend into the bore for the purpose of inspection. The Contractor shall provide all necessary facilities to enable the Engineer to make an inspection of any pile excavation including facilities to check the depth, verticality and position. The full requirements of BS 8008 shall be followed.

The Contractor shall designate a Supervisor to supervise each inspection. The Supervisor shall have a copy of BS 8008 and shall ensure that every person involved in the descent of piles is familiar with its requirements.

### 3.7. Reinforcement

The number of joints in longitudinal steel bars shall be kept to a minimum. Joints in steel reinforcement shall be such that the full strength of each bar is effective across the joint and shall be made so that there is no detrimental displacement of the reinforcement during the construction of the pile, following the guidance of BS 8110. Reinforcement shall be maintained in its correct position during concreting of the pile, to allow a vertical tolerance of +150/−50 mm on the level of the reinforcement projecting above the final cut-off level. Where reinforcement is made up into cages, they shall be sufficiently rigid to enable them to be handled, placed and concreted without damage. If the cage is to be welded together, welding shall be carried out to the requirements of BS 7123. Details of the procedures should be submitted prior to commencement of the works.

Unless otherwise specified, reinforcement shall extend to the base of the pile or to at least 3 m below the bottom of the temporary casing, whichever is the lesser.

Spacers shall be designed and manufactured using durable materials which shall not lead to corrosion of the reinforcement or spalling of the concrete cover. Details of the means by which the Contractor plans to ensure the correct cover to and position of the reinforcement shall be submitted prior to commencement of the works.

## 3.8. Placing concrete
### 3.8.1. General

The workability and method of placing of the concrete shall be such that a continuous monolithic concrete or grout shaft of the full cross-section is formed. Concrete shall be transported from the mixer to the position of the pile in such a manner that segregation of the mix does not occur.

The concrete shall be placed without such interruption as would allow the previously placed batch to have achieved a stiffness which prevents proper amalgamation of the two concrete batches.

The Contractor shall take all precautions in the design of the mix and placing of the concrete to avoid arching of the concrete in a temporary casing. No soil, liquid or other foreign matter shall be permitted to contaminate the concrete.

### 3.8.2. Workability of concrete

Workability measured at the time of discharge into the pile bore shall be in accordance with the limits shown in Table 3.1.

### 3.8.3. Compaction

Internal vibrators shall not be used to compact concrete.

### 3.8.4. Placing concrete in dry borings

Measures shall be taken to ensure that the structural strength of the concrete placed in all piles is not impaired through grout loss, segregation or bleeding.

The method of placing shall be such as to ensure that the concrete or grout in its final position is dense and homogeneous. Concrete shall be introduced into the pile via a hopper and suitable length of rigid delivery tube to ensure that the concrete falls vertically and centrally down the shaft. The tube shall be at least 3 m long.

### 3.8.5. Placing concrete under water or support fluid

Before placing concrete, measures shall be taken in accordance with Clauses 3.6.8 and 3.6.9 to ensure there is no accumulation of silt or other material at the base of the boring, and the Contractor shall ensure that heavily contaminated support fluid, which could impair the free flow of concrete from the tremie pipe, has not accumulated in the bottom of the hole.

A sample of support fluid shall be taken from the top and bottom of the boring using an appropriate device. The fluid shall be prepared, sampled and tested as stated in the Contractor's submission. If tests show the fluid does not comply with the limits stated by the Contractor for the particular type of support fluid, the fluid shall be wholly or partly replaced as appropriate before concrete is placed.

Concrete to be placed under water or support fluid shall be placed by tremie in one continuous operation and shall not be discharged freely into the water or support fluid. Pumping of concrete may be used where appropriate. The bottom end of the tremie must be square to the axis of the tremie and its circumference must be continuous. The tremie must be fully charged with concrete before it is lifted off the base of the pile.

The concrete shall be of high workability in accordance with Table 3.1, Mix C and cement content in accordance with Section 20 and Table 3.1, Mix C.

The concrete shall be placed in such a manner that will not cause segregation of the mix.

The depths to the surface of the concrete shall be measured and

*Specification for piling and embedded retaining walls.* Thomas Telford, London, 1996.

21

the length of the tremie tubes recorded at regular intervals corresponding to the placing of each batch of concrete. The depths measured and volumes placed shall be plotted immediately on a graph and compared with the theoretical relationship of depth against volume.

The hopper and pipe of the tremie shall be clean and watertight throughout. The pipe shall extend to the base of the bore and a sliding plug or barrier shall be placed in the pipe to prevent direct contact between the first charge of concrete in the tremie and the water or support fluid. The pipe shall at all times penetrate the concrete which has previously been placed with a minimum embedment of 3 m and shall not be withdrawn from the concrete until completion of concreting. A sufficient quantity of concrete shall be maintained within the pipe to ensure that the pressure from it exceeds that from the water or support fluid. The internal diameter of the pipe of the tremie shall be of sufficient size to ensure the easy flow of concrete. It shall be so designed that external projections are minimized, allowing the tremie to pass within reinforcing cages without causing damage. The internal face of the pipe of the tremie shall be free from projections.

## 3.9. Extraction of casing
### 3.9.1. Workability of concrete

Temporary casings shall be extracted while the concrete within them remains sufficiently workable to ensure that the concrete is not lifted. During extraction the motion of the casing shall be maintained in an axial direction relative to the pile.

### 3.9.2. Concrete level

When the casing is being extracted, a sufficient quantity of concrete shall be maintained within it to ensure that pressure from external water, support fluid or soil is exceeded and that the pile is neither reduced in section nor contaminated.

The concrete level within a temporary casing in a dry bore may be topped up where necessary during the course of casing extraction so that the base of the casing is always below the concrete surface until the casting of the pile has been completed.

Adequate precautions shall be taken in all cases where excess

Table 3.1.  Piling mix workability

| Piling mix workability | Workability | | Typical conditions of use (The concrete and aggregate sizes must be compatible with the reinforcement spacing) |
|---|---|---|---|
| | Target flow, mm | Slump range, mm | |
| A | not applicable | 75–125 | Placed into water-free unlined or permanently lined bore of 600 mm diameter or over, or where casting level lies below temporary casing; reinforcement spaced at 100 mm centres or greater, leaving ample room for free movement of concrete between bars. |
| B | Target flow 500 +/–50 | Slump* 100–175 | Where reinforcement is spaced at less than 100 mm centres; where cut-off level of concrete is within temporary casing; where pile bore is water-free and the diameter is less than 600 mm. |
| C | Target flow 550 +/–50 | Slump* 150 or more | Where concrete is to be placed by tremie under water or support fluid or by pumping or by continuous flight auger. |

*The slump test method is not suited to these very high workabilities and the flow test is to be preferred.

heads of water or support fluid could occur as the casing is withdrawn because of the displacement of water or fluid by the concrete as it flows into its final position against the walls of the pile bore. Where two or more discontinuous lengths of casing (double casing) are used in the construction the Contractor's method of working shall produce piles to their full designed cross-sections.

The depth to the average levels of the concrete surface of the pile shall be measured before and after each temporary casing is removed. These measurements shall be recorded on the pile record.

### 3.9.3. Pile head casting level tolerances

For piles cast in dry bores using temporary casing and without the use of a permanent casing, piles heads shall be cast to a level above the specified cut-off so that, after trimming, a sound concrete connection with the pile can be made. The casting level shall be within the tolerance above the cut-off level shown in Table 3.2, but shall not be above the commencing surface level. No pile shall be cast with its cut-off level below standing water level unless appropriate measures are taken to prevent inflow of water causing segregation of the concrete as temporary casing is extracted.

For piles cast in dry bores where permanent lining tubes or permanent casings are used, to where their cut-off levels are in stable ground below the base of any casing used, pile heads shall be cast to a level above the specified cut-off so that, after trimming, a sound concrete connection with the pile can be made. The casting level shall be within the tolerance above the cut-off level shown in Table 3.2, but shall not be above the commencing surface level.

For piles cast under water or support fluid, the pile heads shall be cast to a level above the specified cut-off so that, after trimming to remove all debris and contaminated concrete, a sound concrete connection with the pile can be made. The casting level shall be within the tolerance above the cut-off level shown in Table 3.2, but shall not be above the commencing surface level. Cut-off levels may be specified below the standing groundwater level, and where this condition applies the borehole fluid level shall not be reduced below the standing groundwater level until the concrete has set.

Where either support fluid or water is mixed in the ground by the drilling equipment to assist with the installation of temporary casings the casting level shall be coincident with the commencing surface.

*Table 3.2. Casting tolerance above cut-off levels for specified conditions*

| Cut-off level below commencing surface, $H$, m* | Casting tolerance above cut-off level, m | Condition |
|---|---|---|
| 0.15 to any depth | $0.3 + H/10$ | Piles cast in dry bore within permanent casing or cut-off level in stable ground below base of casing |
| 0.15–10.00 | $0.3 + H/12 + C/8$ | Piles cast in dry bore using temporary casing other than above |
| 0.15–10.00 | $1.0 + H/12 + C/8$<br>Where $C$ = length of temporary casing below the commencing surface | Pile cast under water or support fluid** |

\*   Beyond $H$ = 10 m, the casting tolerance applying to $H$ = 10 m shall apply.
\*\*   In cases where a pile is cast so that the cut-off level is within a permanent lining tube, the appropriate tolerance is given by deleting the casing term $C/8$.

### 3.9.4. Temporary backfilling above pile casting level

After each pile has been cast, any empty bore remaining shall be carefully backfilled as soon as possible with inert spoil.

### 3.10. Cutting off pile heads

When cutting off and trimming piles to the specified cut-off level, the Contractor shall take care to avoid shattering or otherwise damaging the rest of the pile. Any cracked or defective concrete shall be cut away and the pile repaired in a manner to provide a full and sound section at the cut-off level.

### 3.11. Grout

The use of grout containing fine aggregate is permitted in place of concrete. The requirements of Sub-section 20.14 shall apply. The fine aggregate shall be in accordance with the limits of grading M or C given in Table 5 of BS 882.

The moisture content of aggregates shall be measured immediately before mixing and as frequently thereafter as is necessary to maintain consistency of the mix. Allowance shall be made for the aggregate moisture when assessing batch weights.

Where the word 'concrete' is used elsewhere in this Section then 'grout' can be read in its place. In particular Sub-section 3.8 shall also apply to grout, except the requirement for grout workability shall be that in Clause 20.14.

### 3.12. Pressure grouting
### 3.12.1. Grouting of piles

Where bases or sides of piles are to be pressure grouted, as detailed in the Particular Specification, the Contractor shall construct the piles with grout tubes and any other necessary equipment pre-installed so that piles may subsequently be grouted.

### 3.12.2. Method of grouting

The method of grouting shall be such that the completed pile meets the requirements of the Specification for load–settlement behaviour and that during grouting pile uplift is within the limits specified. The Contractor shall submit full descriptions of the equipment, materials and methods that he plans to use. The method statement shall comprise at least the following information:

(i) details of specialist Contractor for grouting (if applicable), names of key personnel and their curricula vitae and previous experience on similar types of work
(ii) details of grout pump, mixer, agitator and any other equipment used for mixing and injection of grout
(iii) full details of grout to be used, including additives
(iv) method of quality control on grout, including details of number of cubes taken and checks on density, flow and bleed of the grout
(v) method of measuring grout take, which should be automatic and include a physical method of checking grout take at the end of injection of each circuit
(vi) method of measuring grout pressures which should include a continuous record; calibration certificates for pressure gauges
(vii) typical record sheet for grouting, which shall include records of grout take, grout pressure, residual pressure, times of grouting and pile uplift for each grouting circuit; typical continuous records of grout pressure and pile uplift shall also be included

*(viii)* target minimum, maximum and residual grout pressures for each grout injection

*(ix)* method of grout injection, including full details of any packers; target grout volumes for each injection

*(x)* method of measuring friction losses in the tubes/packers.

### 3.12.3. Grout tubes

The grouting tubes shall be tested to determine any grout leakage in joints under pressure prior to installation into the piles. The grout tubes must be capable of withstanding the pressures to which they will be subjected.

Robust threaded caps shall be provided to protect the top of the grouting tubes during concreting and afterwards.

The grouting tubes shall be flushed with water after each grouting operation. If the target grout pressures are not achieved, or the specified uplift is not achieved, then the pile shall be regrouted within 24 hours.

The Contractor shall provide an engineer to monitor grout pressures and grout takes.

### 3.12.4. Pile uplift

During base grouting, the pile uplift shall be not less than 0.2 mm and shall not exceed 2.0 mm.

The Contractor shall provide an engineer to monitor the uplift of the piles during grouting.

The pile uplift shall be monitored with an appropriate level reading off a graduated scale attached to the pile head and by a dial gauge attached to a reference frame. A continuous record of pile uplift shall be made using a displacement transducer also attached to the reference frame. The accuracy of measurement shall be a minimum of 0.1 mm for each monitoring device. The design of the reference frame shall be submitted to the Engineer.

### 3.12.5. Grout testing

Close control of the mixing of the grout shall be carried out. The Contractor shall provide and maintain on site all test facilities required to test and control the grout mixes.

### 3.12.6. Records

The Contractor shall provide duplicate copies of all grouting records for each pile within 24 hours of the completion of grouting that pile.

The records shall comprise the following information:

*(a)* pile number

*(b)* date

*(c)* leakage test on grout pipes

*(d)* grout mix

*(e)* continuous records of grout pressure and pile uplift to the same timescale. The records shall be annotated at regular intervals with physical measurements from pressure gauges and dial gauges including times of such measurements. Circuit number and pile number are also to be included.

*(f)* for each circuit the peak and residual grout pressures including times and sequence of grouting

*(g)* the total pile uplift and the uplift after each circuit is grouted

*(h)* for each circuit the physical and automatic measurements for grout volume injected and derived injected volumes; the total grout volume

*(i)* all tests made on grout.

# 4. Bored piles constructed using continuous flight augers and concrete or grout injection through hollow auger stems

## 4.1. General

All materials and work shall be in accordance with Sections 1, 4 and 20 of this Specification, except where there may be conflict of requirements, in which case this Section shall take precedence. The abbreviation 'cfa' can be used to describe these piles.

## 4.2. Particular Specification

The following matters are, where appropriate, described in the Particular Specification:

(a) specified working loads
(b) additional performance criteria for piles under test (also see Table 1.3)
(c) sampling and testing of pile materials
(d) types of cement
(e) cement replacement materials
(f) concrete or grout admixtures
(g) types and sizes of aggregate
(h) grades of concrete or grout
(i) method of testing concrete or grout workability
(j) designed concrete or grout mixes and maximum free water to cement ratio
(k) grades and types of and cover to reinforcement
(l) pile dimensions
(m) preliminary piles
(n) trial bores
(o) disposal of excavated material
(p) other particular requirements.

## 4.3. Construction of pile

The completed pile shaft shall consist of a continuous circular concrete column with a minimum diameter at least equal to the nominal pile diameter. The concreting process shall produce a concrete column of uniform quality free from bleeding and segregation. Inclusions of soil or other debris within the concrete area are not admissible.

The pile shall be bored using suitable equipment capable of penetrating the ground without drawing surrounding soils laterally into the pile bore.

Concrete shall be delivered to the pile through suitable tubing and the hollow auger stem. All pipe fitments and connections shall be so constructed that concrete does not leak during the injection process.

The concrete shall be applied to the pile at a sufficient rate during auger withdrawal to ensure that a continuous monolithic shaft of the full specified cross-section is formed, free from debris or any segregated ground.

At the beginning of concrete placing this sealing device shall be removed by the application of concrete pressure. Care shall be taken to ensure that the auger is lifted only sufficiently to initiate the flow of concrete, and that water inflow and soil movement at the base of the auger are prevented. The technique and equipment used to initiate and maintain the concrete flow shall be such that a pile of the full specified cross-section is obtained from the maximum depth of boring to the final pile cut-off level.

### 4.4. Setting out

The position and verticality of the auger immediately prior to the construction of each pile shall be checked by the Contractor. Any checks by the Engineer shall not relieve the Contractor of his responsibility.

### 4.5. Diameter of piles

The diameter of a pile shall be not less than the specified diameter.

The cutting head width shall be checked as necessary and recorded for each pile to ensure the specified diameter is achieved and that the width is greater than the diameter of the following flight. A tolerance of 5% on the cutting head width is permissible.

### 4.6. Boring
#### 4.6.1. Boring near recently cast piles

Piles shall be formed in an order and in such a manner that no damage is sustained by previously formed piles.

Piles shall not be bored so close to other piles which have recently been cast and which contain workable or unset concrete such that a flow of concrete could be induced from or damage caused to any of the piles.

#### 4.6.2. Removal of augers from the ground

The auger shall not be extracted from the ground during the boring or construction of a pile in such a way that an open unsupported bore or inflow of water into the pile section would result.

If during the augering of a continuous flight auger injection pile it is necessary to raise the auger and subsequently to re-auger, the required depth shall be increased to at least 0.5 m below the depth previously reached if this is practical, and the fact shall be recorded on the pile record. As the auger is withdrawn from the ground it shall be cleaned of all rising soil.

#### 4.6.3. Suitability of boring equipment

The piles shall be bored using equipment capable of penetrating the ground without drawing surrounding soils laterally into the pile bore.

The Contractor shall record the fact if flighting of soil up the auger is excessive.

The verticality of the auger shall be checked at the commencement of boring. Should it deviate during boring so that the pile verticality is outside the specified tolerance the fact shall be recorded on the pile record.

Lengths of auger shall not be joined together during boring nor split during auger extraction.

#### 4.6.4. Sealing the base of the auger

The base of the auger stem shall be fitted with a suitable means of sealing it against ingress of water and soil during boring.

### 4.7. Depth of piles

Any failure of a pile to reach the required depth, as given in the Particular Specification, shall be reported to the Engineer without delay and a full statement of the reasons given.

### 4.8. Placing of concrete or grout
#### 4.8.1. Concrete mix design and workability

Where not otherwise stated in this section, the concrete shall comply with Section 20 of this Specification. The design and workability of concrete to be used in the formation of a pile shall produce a mix which is suitable for pumping. It shall have a slump range of 150 mm or greater or target flow of 550 mm with tolerance

*Specification for piling and embedded retaining walls.* Thomas Telford, London, 1996.

27

+/−50 mm and a minimum cement content of 340 kg/m$^3$. The fine aggregate shall be in accordance with the limits of grading M or C given in Table 5 of BS 882. The mix shall be designed so that segregation does not occur during the placing process, and bleeding of the mix shall be minimized.

The workability of concrete mixes shall be measured by the method described in the Particular Specification.

### 4.8.2. Grout mix design and workability

Where not otherwise stated in this section, the grout shall comply with Sub-section 20.14, with the exception that fine aggregate may be used. The fine aggregate shall be in accordance with the limits of Grading M or C given in Table 5 of BS 882.

The moisture content of aggregates shall be measured immediately before mixing and as frequently thereafter as is necessary to maintain consistency of the mix. Allowance shall be made for the aggregate moisture content when assessing batch weights.

The procedure for monitoring the suitability of each batch of grout throughout the Works shall be notified to the Engineer prior to commencement of the Works and shall include the measurement of flow, bleed and density.

### 4.8.3. Equipment for supply of concrete or grout to piles

Grout or concrete shall be supplied to the pile through suitable concrete pump, tubing and the hollow auger stem. All pipe fitments and connections shall be so constructed that grout does not leak during the injection process.

### 4.8.4. Commencement of concrete or grout supply to each pile

At the beginning of concrete or grout placing the sealing device at the base of the auger stem shall be removed by the application of a positive concrete or grout pressure at the point of measurement. Care shall be taken to ensure that the auger is lifted only the minimum distance necessary to initiate the flow of concrete or grout, and that water inflow and soil movement at the base of the auger are minimized. The technique and equipment used to initiate and maintain the concrete or grout flow shall be such that a pile of the full specified cross-section is obtained from the maximum depth of boring to the final pile cut-off level.

### 4.8.5. Rate of supply of concrete or grout

The concrete or grout shall be supplied to the pile at a sufficient rate during auger withdrawal to ensure that a continuous monolithic shaft of at least the full specified cross-section is formed, free from debris or any segregated concrete or grout.

If rotation of the auger occurs during auger extraction, it shall be positive, i.e. in the direction that would cause the auger to penetrate into the concrete.

### 4.8.6. Completion of concreting or grouting of piles

If the concrete or grout placing in any pile cannot be completed in the normal manner, then the pile shall be rebored to a safe level below the position of interruption of supply before concrete or grout has achieved initial set and before further concrete or grout is injected, if this is possible, and the fact shall be recorded on the pile record. The method statement shall set out the procedure.

### 4.8.7. Casting level of pile head

Concrete or grout shall be cast to the commencing surface level in all cases. The pile position shall be clearly marked and fenced off so as not to cause a safety hazard.

### 4.9. Reinforcement

All reinforcement shall be placed with the minimum delay after the completion of the concreting or grouting operation. It shall be fabricated in cages or bundles of bars fixed securely to permit it to be placed in the correct position and to the depth specified through the concrete or grout of the pile. Suitable spacers shall be provided to maintain the specified concrete or grout cover to steel.

The transverse reinforcement of any reinforcing cage shall meet the design requirements and shall maintain the longitudinal bars in position when the cage is inserted into the wet concrete or grout.

Longitudinal main steel reinforcement shall be continuous over the specified length. Where joints are necessary, no more than one joint shall be used and then only if the reinforcement cage length exceeds 12 m. Joints in steel reinforcement shall be such that the full strength of each bar is effective across the joint and shall be made so that there is no detrimental displacement of the reinforcement during placing in the pile, following the guidance of BS 8110. At the joint bars shall be welded to the requirements of BS 7123 or joined together in a suitable manner.

Reinforcement shall be placed and maintained in position to provide the specified projection of reinforcement above the final cut-off level. A vertical tolerance of +150/−50 mm on the level of reinforcement projecting above the final cut-off level shall be met.

The Contractor shall provide such additional reinforcement as he may require to suit his method of placing the reinforcing steel cage.

## 4.10. Monitoring system for pile construction

An automated system shall be provided for monitoring the construction of the piles. It shall provide the operator with information on the depth of the auger tip, flow of concrete and relative pressure of concrete as a minimum. In addition the automatic monitoring equipment shall monitor continuously with depth the following parameters:

(*a*) during boring:
   (*i*) auger penetration rate

(*b*) during concreting:
   (*i*) rate of extraction of the auger
   (*ii*) relative injection pressure of concrete or grout
   (*iii*) rate of supply of concrete or grout.

Equipment used for monitoring shall be calibrated at the start of the piling works and calibration certificates issued to the Engineer prior to commencement of the piling. After the commencement of the works, the monitoring equipment shall be calibrated at the frequency specified below or at any time when there is reason to suspect malfunction.

Depth shall be calibrated once a week. At full auger length the tolerance is ±0.1 m.

Concrete flow rate shall be calibrated once during progress of the works by passing a known volume of concrete through the flowmeter. The tolerance on flow is ±5%.

The pressure transducer shall have a calibration certificate. The tolerance on pressure shall be stated in bars.

The rig operator shall be competent and experienced in the construction of continuous flight auger piles and details of relevant experience shall be submitted prior to work commencing. A full-time supervisor shall be devoted to pile construction and his relevant experience shall also be submitted prior to work commencing.

The automated monitoring system must be operational at the start of every pile.

The following information shall be recorded during the construction of each pile:

(*i*)  the incremental time of auger penetration during boring
(*ii*)  final depth of the bore
(*iii*)  time for each 0.5 m increment of auger extraction during concreting
(*iv*)  the volume of concrete or grout pumped for each 0.5 m increment of auger extraction during concreting
(*v*)  the total volume of concrete for completion of the pile
(*vi*)  the time at the start and end of boring, concreting and insertion of the reinforcement cage and any time intervals for delays or stoppages.
(*vii*)  relative concrete pressure at the top of the hollow auger stem during concreting
(*viii*) the direction of rotation of the auger during concreting.

If the number of auger revolutions relative to auger penetration becomes abnormally high at any stage during boring, the fact shall be recorded, as shall the occurrence of excessive flighting.

To facilitate manual monitoring of pile construction the following shall be provided:

(*i*)  the mast of the piling rig shall have paint marks at 0.25 m intervals over its full height and the metre increments shall be marked numerically
(*ii*)  an automatic stroke counter on the concrete pump to record the number of piston strokes.

The monitoring results must be made available to the Engineer immediately on completion of every pile.

The Contractor shall submit proposals on how he shall complete a pile in the event of failure of all or part of the rig instrumentation system prior to the commencement of work. This proposal shall include the recording of the depth at which failure occurred, the time for auger extraction during concreting, and the total volume of concrete or grout delivered. Any pile which was subject to an instrumentation failure during construction shall be integrity tested in accordance with Section 9.

## 4.11. Cutting off pile heads

When cutting off and trimming piles to the specified cut-off level, the Contractor shall take care to avoid shattering or otherwise damaging the rest of the pile. Any laitance, or contaminated, cracked or defective concrete shall be cut away and the pile made good in a manner to provide a full and sound section up to the cut-off level.

*Specification for piling and embedded retaining walls.* Thomas Telford, London, 1996.

# 5. Driven cast-in-place piles

### 5.1. General

All materials and work shall be in accordance with Sections 1, 5 and 20 of this Specification, except where there may be conflict of requirements, in which case this Section shall take precedence.

### 5.2. Particular Specification

The following matters are, where appropriate, described in the Particular Specification:

(a) specified working loads
(b) additional performance criteria for piles under test (also see Table 1.3)
(c) sampling and testing of pile materials
(d) type of cement
(e) cement replacement materials
(f) types and sizes of aggregate
(g) grades of concrete
(h) method of testing concrete workability
(i) designed or prescribed mixes and maximum free water to cement ratio
(j) grades and types of and cover to reinforcement
(k) types and quality of permanent casing
(l) types and quality of pile shoes (where required)
(m) penetration or depth or toe level
(n) driving resistance or dynamic evaluation or set
(o) trial drives
(p) preliminary piles
(q) uplift/lateral displacement trials
(r) preboring or jetting or other means of easing pile driveability
(s) detailed requirements for driving records (including requirements for measurement of temporary compressions and redrives)
(t) disposal of excavated material and cut-off heads of piles
(u) other particular requirements.

### 5.3. Materials
#### 5.3.1. Permanent casings

Permanent casings shall be as specified in the Particular Specification. Where a permanent casing is to be made from a series of short sections it shall be watertight. The dimensions and quality of the casing shall be adequate to withstand the stresses caused by handling and driving without damage or distortion.

#### 5.3.2. Pile shoes

Pile shoes shall be manufactured from durable material capable of withstanding the stresses caused by driving without damage, and shall be designed to give a watertight joint during construction.

### 5.4. Diameter of piles

The diameter of a pile shall be not less than the specified diameter.

### 5.5. Temporary casings

Temporary casings shall be free from significant distortion. They shall be of uniform external cross-section throughout each continuous length and shall be of sufficient strength to withstand driving and ground forces without deformation. During concreting they shall be free from internal projections and encrusted concrete which might prevent the proper formation of piles.

## 5.6. Enlarged pile bases

Where the Contractor wishes to form a pile with an enlarged base or where such a base is specified in the Particular Specification, details of his method of forming the base and the materials to be used shall be submitted with the Contractor's method statement.

## 5.7. Driving piles
### 5.7.1. Piling near recently cast piles

Casings shall not be driven or piles formed so close to other piles which have recently been cast and which contain workable or unset concrete that a flow of concrete could be induced from or significant damage caused to any of the piles.

### 5.7.2. Performance of driving equipment

The Contractor shall provide the Engineer with information on the efficiency and energy of the driving equipment including when followers are used. Where required in the Particular Specification, dynamic evaluation and analysis shall be provided.

Drop hammers shall not be used from floating craft in such a manner as to cause instability of the craft or significant damage to the pile.

### 5.7.3. Length of piles

The length of pile to be driven in any position shall comply with the Contractor's schedule, unless otherwise specified in the Particular Specification. During the execution of the Works, any amendments to the scheduled lengths shall be reported to the Engineer giving adequate notice prior to installation of the piles concerned.

### 5.7.4. Driving procedure and redrive checks

The driving procedure shall be such so as to avoid damage to the piles.

The Engineer shall be given 24 hours' notice of commencement of the driving of the first pile.

The Contractor shall give adequate notice and provide all facilities to enable the Engineer to check driving resistance, sets and temporary compressions.

Each pile casing shall be driven continuously until the depth or set as required by the design has been reached. In the event of unavoidable interruption to driving, the pile may be redriven provided the casing can subsequently be driven to the specified depth and/or resistance or set without damage.

Driving records shall be made for every pile. This record shall contain the weight and fall of the hammer or ram and the number of blows for each 0.25 m of penetration, unless otherwise specified in the Particular Specification.

The Contractor shall inform the Engineer without delay if an unexpected change in driving characteristics is encountered.

### 5.7.5. Set

The set and temporary compression shall be measured and recorded for each pile at the completion of driving unless otherwise stated in the Particular Specification.

When a set is being measured, the following requirements shall be met.

(a) The exposed part of the piles casing shall be in good condition, without damage or distortion.

(b) The dolly, helmet and packing, if any, shall be in sound condition.

(c) The hammer blow shall be in line with the pile axis and the impact surfaces shall be flat and at right angles to the pile and hammer axis.

*Specification for piling and embedded retaining walls.* Thomas Telford, London, 1996.

(d) The hammer shall be in good condition, delivering adequate energy per blow, and operating correctly.

(e) Temporary compression of the pile casing shall be recorded, if required in the Particular Specification.

The set shall be recorded either as the penetration in millimetres per 10 blows or as the number of blows required to produce a penetration of 25 mm.

### 5.7.6. Driving sequence and risen piles

Piles shall be driven in a sequence to minimize any detrimental effects of heave and lateral displacement of the ground.

The sequence and method of piling including preboring shall limit uplift and lateral movement so that the final position of each pile is within the specified tolerances. At all times the deflections of each pile from its axis as formed shall not be such as to cause damage or impair durability of the piles or any structures or services.

The maximum permitted uplift of each pile due to any one pile driven within a pile centre to centre radius of eight pile diameters is 3 mm unless it can be demonstrated by static load testing that uplift exceeding this amount does not affect the ability of the pile to meet the requirements of this Specification.

Even if during the installation of preliminary piles uplift is shown to be within the permitted maximum and the preliminary piles tested meet the requirements of the Specification checks of uplift on working piles shall be made by the Contractor at least once a week throughout the period of the piling work and the results reported to the Engineer.

If preliminary piles are not installed the Contractor shall commence installation of working piles taking measures to reduce or eliminate uplift until it can be established by site measurements that such measures are no longer necessary. Thereafter checks on uplift shall be made by the Contractor at least once a week and the results reported to the Engineer.

If a static load test shows that a pile which is uplifted more than the maximum permitted amount due to any pile driven within a radius of eight pile diameters does not comply with the requirements of the Specification the Contractor shall submit details of remedial works for all such piles that may have been uplifted.

If records and measurements show that piles have been laterally displaced so as to be outside the permitted tolerance, damaged or are of impaired durability the measures the Contractor plans to adopt to enable the piles to comply with the Specification shall be provided to the Engineer.

Laterally displaced piles shall not be corrected by forceable correction at the heads.

### 5.7.7. Preboring

If preboring is specified the pile casing shall be pitched after preboring to the depth and diameter stipulated in the Particular Specification.

Other means to ease pile driveability may be used provided the completed piles meet their specified requirements.

### 5.7.8. Internal drop hammer

Where a casing for a pile without an enlarged base is to be driven by an internal drop hammer, a plug consisting of concrete grade 20 with a water/cement ratio not exceeding 0.25 shall be placed in the pile. This plug shall have a compacted height of not less than $2\frac{1}{2}$ times the diameter of the pile. Fresh concrete shall be added to ensure that this height of driving plug is maintained in the casing throughout the period of driving, and in any event a plug of fresh concrete shall be added after $1\frac{1}{2}$ hours of normal driving or after 45 minutes of hard driving, or, should the pile-driving be interrupted for 30 minutes or longer, fresh concrete shall be added prior to driving being resumed.

## 5.8. Repair of damaged pile heads and extending of piles to the cut-off level

When repairing or extending the head of a pile, the head shall be cut off square at sound concrete, and all loose particles shall be removed by wire-brushing, followed by washing with water.

If the level of sound concrete of the pile is below the cut-off level, the pile shall be extended to the cut-off level with concrete of a grade not inferior to that of the concrete of the pile so that it will safely withstand the specified working load.

## 5.9. Lengthening of permanent pile casings during construction

The lengthening of permanent steel pile casings by adding an additional length of steel casing during construction shall be carried out in accordance with the relevant clauses of Section 6 of this Specification.

## 5.10. Inspection and remedial work

Prior to placing concrete in a pile casing, the Contractor shall check that the casing is undamaged, and free from water or other foreign matter. In the event of water or foreign matter having entered the pile casing, the casing shall be withdrawn, repaired if necessary and redriven, or other action taken to continue the construction of the pile to meet the requirements of the Specification.

## 5.11. Reinforcement

The number of joints in longitudinal steel bars shall be kept to a minimum. The full strength of each bar shall be effective across each joint, which shall be made so that there is no detrimental displacement of the reinforcement during the construction of the pile, following the guidance of BS 8110. Reinforcement shall be maintained in its correct position during concreting of the pile to allow a vertical tolerance of +150/−50 mm on the level of the reinforcement projecting above the final cut-off level to be met. Where it is made up into cages, they shall be sufficiently rigid to enable them to be placed, handled and concreted without damage. If the cage is to be welded together, welding shall be carried out to the requirements of BS 7123.

Spacers shall be designed and manufactured using durable materials which shall not lead to corrosion of the reinforcement or spalling of the concrete cover. Details of the means by which the Contractor plans to ensure the correct cover to and position of the reinforcement shall be submitted.

## 5.12. Placing concrete
### 5.12.1. General

The workability and method of placing of the concrete shall be such that a continuous monolithic concrete shaft of the full cross-section is formed. Concrete shall be transported from the mixer to the position of the pile in such a manner that segregation of the mix does not occur.

The concrete shall be placed without such interruption as would allow the previously placed batch to have hardened.

The Contractor shall take all precautions in the design of the mix and placing of the concrete to avoid arching of the concrete in the casing. No spoil, liquid or other foreign matter shall be permitted to contaminate the concrete.

*Specification for piling and embedded retaining walls.* Thomas Telford, London, 1996.

### 5.12.2. Workability of concrete

Workability measured at the time of discharge into the pile casing shall be in accordance with the requirement shown in Table 3.1, except that these standards shall not apply to piling systems which use semi-dry concrete and employ special means for its compaction. The concrete shall be of the workability specified when in its final position and until all construction procedures in forming the pile have been completed.

### 5.12.3. Compaction

Internal vibrators shall not be used to compact concrete cast-in-place.

### 5.12.4. Placing concrete

Measures shall be taken as necessary in all piles to ensure that the structural strength of the placed concrete is not impaired through grout loss, segregation or bleeding.

### 5.13. Extraction of casing
### 5.13.1. Workability of concrete

Temporary casings shall be extracted while the concrete within them remains sufficiently workable to ensure that the concrete is not lifted. Should a semi-dry mix have been used, the Contractor shall ensure the concrete shall not lift during extraction of the casing.

### 5.13.2. Concrete level

When the casing is being extracted, a sufficient quantity of concrete shall be maintained within it to ensure that pressure from external water or soil is exceeded and that the pile is neither reduced in section nor contaminated.

Concrete shall be topped up as necessary while the casing is extracted until the required head of concrete to complete the pile in a sound and proper manner has been provided. No concrete is to be placed once the bottom of the casing has been lifted above the top of the concrete.

### 5.13.3. Vibrating extractors

The use of vibrating casing extractors will be permitted subject to compliance with Clause 1.12.

### 5.13.4. Concrete casting tolerances

For piles constructed without the use of a rigid permanent lining, pile concrete shall be cast to the commencing surface level.

Where piles are constructed inside rigid permanent lining tubes or permanent casings, pile heads shall be cast to a level above the specified cut-off so that, after trimming, a sound concrete connection with the pile can be made. In this case the tolerance of casting above the cut-off level shall be determined according to Table 3.2.

### 5.14. Temporary backfilling above pile casting level

After each pile has been cast, any hole remaining shall be protected and shall be carefully backfilled as soon as possible with appropriate materials.

### 5.15. Cutting off pile heads

When cutting off and trimming piles to the specified cut-off level, the Contractor shall take care to avoid shattering or otherwise damaging the rest of the pile. Any cracked or defective concrete shall be cut away and the pile repaired to provide a full and sound section to the cut-off level.

# 6. Steel bearing piles

## 6.1. General

All materials and work shall be in accordance with Sections 1 and 6 of this Specification, except where there may be conflict of requirements, in which this Section shall take precedence.

## 6.2. Particular Specification

The following matters are, where appropriate, described in the Particular Specification:

    (*a*)  specified working loads

    (*b*)  additional performance criteria for piles under test (also see Table 1.3)

    (*c*)  grades of steel

    (*d*)  sections of proprietary types of pile

    (*e*)  thickness of circumferential weld reinforcement

    (*f*)  minimum length of pile to be supplied

    (*g*)  types of head and toe preparation

    (*h*)  types of pile shoe (where required)

    (*i*)  surface preparation

    (*j*)  types of coating

    (*k*)  thickness of primer and coats

    (*l*)  welding procedure; additional requirements to Sub-section 6.6

    (*m*)  non-destructive testing of welds; additional requirements to Clauses 6.7.3 and 6.7.4

    (*n*)  concreting of piles

    (*o*)  penetration or depth or toe level

    (*p*)  driving resistance or dynamic evaluation or set

    (*q*)  trial drives

    (*r*)  preliminary piles

    (*s*)  uplift/lateral displacement trials

    (*t*)  preboring or jetting or other means of easing pile driveability

    (*u*)  detailed requirements for driving records (including requirements for measurement of temporary compressions and redrives)

    (*v*)  disposal of cut-off heads of piles

    (*w*)  other particular requirements

## 6.3. Ordering of piles

The Contractor shall ensure that the piles are available at the time for incorporation in the works. All piles and production facilities shall be made available for inspection at any time. Piles shall be carefully examined at the time of delivery and any faulty piles replaced. Records of the examination shall be provided to the Engineer. The records of testing of the steel used in the piles shall be made available to the Engineer. The Contractor shall submit details of all preliminary test results at least five working days before piles for the main work are ordered.

## 6.4. Materials
### 6.4.1. Pile shoes

Cast steel shoes shall be of steel to BS 3100, Grade A1, flat plate and welded fabricated steel shoes shall be to BS 4360 Grade 43A or 50A or to BS EN10 025 Grade Fe 430A or Fe 510A.

### 6.4.2. Strengthening of piles

The strengthening to the toe of a pile in lieu of a shoe or the strengthening of the head of a pile shall be made using material of the same grade as the pile.

### 6.4.3. Manufacturing tolerances

All piles shall be of the type and cross-sectional dimensions as designed. For standard rolled sections the dimensional tolerances and weight shall comply with the relevant standard. Length tolerance of H-section steel bearing piles shall be ±50 mm in accordance with BS EN10 034. The tolerance on length shall be −0 and +75 mm unless otherwise specified. For proprietary sections the dimensional tolerances shall comply with the manufacturer's standards. The rolling or manufacturing tolerances for proprietary sections shall be such that the actual weight of section does not differ from the theoretical weight by more than +4% or −2.5%. The rolling or manufacturing tolerances for steel tubular piles shall be such that the actual weight of section does not differ from the theoretical weight by more than +5% or −5%. The rolling or proprietary tolerances for H-section steel bearing piles shall be such that the actual weight of the section does not differ from the theoretical weight by more than +2.5% or −2.5%.

### 6.4.4. Straightness of sections

For standard rolled sections the deviation from straightness shall be within the compliance provisions of BS EN10 034. When two or more rolled sections are joined by butt-jointing, the deviation from straightness shall not exceed 1/600 of the overall length of the pile.

For proprietary sections made up from rolled sections and for tubular piles, the deviation from straightness on any longitudinal face shall not exceed 1/600 of the length of the pile nor 5 mm in any 3 m length.

### 6.4.5. Fabrication of piles

For tubular piles where the load will be carried by the wall of the pile, and if the pile will be subject to loads that induce reversal of stress during or after construction, the external diameter at any section as measured by using a steel tape on the circumference shall not differ from the theoretical diameter by more than +1% or −1%.

The ends of all tubular piles as manufactured shall be within a tolerance on ovality of +1% or −1% as measured by a ring gauge for a distance of 100 mm at each end of the pile length.

The root edges or root faces of lengths of piles that are to be shop butt-welded shall not differ by more than 25% of the thickness of pile walls not exceeding 12 mm thick or by more than 3 mm for piles where the wall is thicker than 12 mm. When piles of unequal wall thickness are to be butt-welded, the thickness of the thinner material shall be the criterion.

Pile lengths shall be set up so that the differences in dimensions are matched as evenly as possible.

### 6.4.6. Matching of pile lengths

Longitudinal seam welds and spiral seam welds of lengths of tubular piles forming a complete pile shall whenever possible be evenly staggered at the butt, but if, in order to obtain a satisfactory match of the ends of piles or the specified straightness, the longitudinal seams or spiral seams are brought closely to one alignment at the joint then they shall be staggered by at least 100 mm.

### 6.4.7. Welding

All welding shall be to BS 5135. For a tubular pile where the load will be compressive and non-reversible and will be carried by the wall of the pile or by a concrete core, the welding shall be to BS 5135 or BS 6265.

### 6.4.8. Inspection and test certificates

The Contractor shall provide the Engineer with works test certificates, analyses, and mill sheets, together with a tube manufacturer's certificate showing details of the pipe number, cast number of the steel and a record of all tests and inspections carried out.

The Contractor shall give the Engineer adequate notice of the start of each stage of the manufacturing process and any production tests and shall make facilities available for the Engineer to inspect. The Contractor shall provide the Engineer with samples when required.

### 6.5. Welders' qualifications

Only welders who are qualified to BS EN287, Part 1, and have a proven record over the previous six months, or who have attained a similar standard, shall be employed on the Works. Proof of welders' proficiency shall be made available to the Engineer on request.

### 6.6. Welding procedures

All welding procedures shall have been qualified to BS EN288 and the Contractor shall submit full details of the welding procedures and electrodes, with drawings and schedules as may be necessary. Tests shall be undertaken as may be required by the relevant British Standard or as may be required in the Particular Specification.

### 6.7. Manufacturing processes
### 6.7.1. Welded tube piles

The Engineer shall be informed if different edge preparation from that shown on the drawings is required for use with automatic welding machines or because of the method of rolling.

The manufacturer shall submit details of the manufacturing and welding procedures before commencement of manufacture to the Engineer.

### 6.7.2. Welded box piles and proprietary sections

Welded box piles or proprietary sections made up from two or more hot-rolled sections shall be welded in accordance with the manufacturer's standards.

### 6.7.3. Non-destructive testing of welds

During production of welded tube piles, one radiograph or ultrasonic test of a length of approximately 300 mm shall be made at each end of a length as manufactured at the mill to provide a spot check on weld quality. In addition, on a spirally welded tube pile, a further check shall be made on welded joints between strip lengths.

All the circumferential welds shall be fully radiographed or ultrasonically tested by the method specified in the Particular Specification.

Results shall be submitted to the Engineer within 10 days of the completion of the tests. If the results of any weld test do not conform to the specified requirements, two additional specimens from the same length of pile shall be tested. In the case of failure of one or both of these additional tests, the length of pile covered by the tests shall be rejected.

### 6.7.4. Standards for welds

*Longitudinal welds in tubular piles.* For piles of longitudinal or spiral weld manufacture where the load will be carried by the wall of the pile, and if the pile will be subject to loads which induce reversal of stress during or after construction other than driving stresses, the standard for interpretation of non-destructive testing shall be the latest edition of the American Petroleum Institute Specification 5L. The maximum permissible height of weld reinforcement shall not exceed 3.2 mm for wall thicknesses not exceeding 12.7 mm and 4.8 mm for wall thicknesses greater than 12.7 mm.

*Longitudinal welds in proprietary box piles and proprietary sections.* Longitudinal welds joining the constituent parts of the box or proprietary section shall be in accordance with the manufacturer's specification.

*Circumferential welds.* For circumferential welds in tubular piles the same maximum height of weld reinforcement as specified above for longitudinal welds in tubular piles shall apply, the standard for interpretation of non-destructive testing shall be the latest edition of American Petroleum Institute Specification 5L.

## 6.8. Site-welded butt splices
### 6.8.1. Support and alignment

When lengths of pile are to be butt-spliced on site, adequate facilities shall be provided for supporting and aligning them prior to welding such that the specified straightness criterion can be achieved.

### 6.8.2. Weld tests

Weld tests shall be performed by radiographic or ultrasonic methods as specified. Provided that satisfactory results are being obtained, one test of a length of 300 mm shall be made for 10% or more of the number of welded splices in the case where the load will be carried by the wall or section of the pile. Where the load will be carried by the concrete core of a pile the number of tests will be decided by the Engineer, but will not normally exceed 10% of the number of butt splices.

Results shall be submitted to the Engineer within 10 days of completion of the tests. Any defective weld shall be cut out, replaced and reinspected.

### 6.8.3. Standards for site butt welds

Welds shall comply with the requirements of the Weld quality standards for site butt welds in steel bearing piles published by British Steel Corporation, General Steels Group.

### 6.8.4. Protection during welding

All work associated with welding shall be protected from the weather so that the quality of work meets the requirements of the Specification.

## 6.9. Coating piles for protection against corrosion
### 6.9.1. Durability of coatings

Where coatings are specified they shall be provided in accordance with the Particular Specification.

### 6.9.2. Definition

The term 'coating' shall include the primer and the coats specified.

### 6.9.3. Specialist labour

The preparation of surfaces and the application of the coats to form the coating shall be carried out by specialist labour having experience in the preparation of the surface and the application of the coating specified.

### 6.9.4. Protection during coating

All work associated with surface preparation and coating shall be undertaken inside a waterproof structure.

### 6.9.5. Surface preparation

All surfaces to be coated shall be clean and dry and prepared by one or both of the following methods, as specified.

Degreasing with detergent wash compatible with the coating shall be carried out where necessary.

All surfaces shall be blast cleaned to Sa 2.5 of BS 7079 Part A1. Blast-cleaning shall be done after fabrication. Unless an instantaneous-recovery blasting machine is used, the cleaned steel surface shall be air-blasted with clean dry air and vacuum-cleaned or otherwise freed from abrasive residues and dust immediately after cleaning.

### 6.9.6. Application and type of primer

Within 4 hours after surface preparation, before visible deterioration takes place, the surface shall be coated with an appropriate primer or the specified coating. No coating shall be applied to a metal surface which is not thoroughly dry.

The primer shall be compatible with the specified coating and shall be such that if subsequent welding or cutting is to be carried out it shall not emit noxious fumes or be detrimental to the welding.

### 6.9.7. Control of humidity during coating

No coating shall be applied when the surface metal temperature is less than 3°C above the dewpoint temperature or when the humidity could have an adverse effect on the coat.

When heating or ventilation is used to secure suitable conditions to allow coating to proceed, care shall be taken to ensure the heating or ventilation of a local surface does not have an adverse effect on adjacent surfaces or work already done.

### 6.9.8. Parts to be welded

The coating within 200 mm of a weld shall be applied after welding. The method of application shall comply with the manufacturer's recommendations.

### 6.9.9. Thickness, number and colour of coats

The nominal thickness of the finished coating and if necessary of each coat shall be as specified. The average coat or finished coating thickness shall be equal to or greater than the specified nominal thickness. In no case shall any coat or finished coating be less than 75% of the nominal thickness. Each coat shall be applied after an interval that ensures the proper hardening or curing of the previous coat.

Where more than one coat is applied to a surface, each coat shall, if possible, be of a different colour from the previous coat. The colour sequence and final coating colour shall be notified.

### 6.9.10. Inspection of coatings

The finished coating shall be generally smooth, of dense and uniform texture and free from sharp protuberances or pin-holes. Excessive sags, dimpling or curtaining shall be retreated.

Any coat damaged by subsequent processes or which has deteriorated to an extent such that proper adhesion of the coating is in doubt shall be removed, and the surface shall be cleaned to the original standard and recoated to provide the specified number of coats.

The completed coating shall be checked for thickness by a magnetic thickness gauge. Areas where the thickness is less than that specified shall receive additional treatment.

The completed coating shall be checked for adhesion by means of an adhesion test to BS 3900 Part E6, carried out on 10% of the piles. The adhesion of any completed coating shall not be worse than Classification 2. Adhesion tests should not be carried out until seven days after coating. On completion of testing the test area shall be made good to the standard specified in the Particular Specification. Areas where the adhesion is defective shall be repaired and reinspected.

### 6.10. Marking, handling and storage of piles
### 6.10.1. Marking of piles

Each pile shall be clearly numbered and its length shown near the pile head using white paint. In addition, before being driven, each pile shall be graduated along its length at intervals of 250 mm.

### 6.10.2. Handling and storage of piles

All piles within a stack shall be in groups of the same length and on appropriate supports. All operations such as handling, transporting and pitching of piles shall be carried out in a manner such that no significant damage occurs to piles and their coatings.

### 6.11. Driving of piles
### 6.11.1. Leaders and trestles

At all stages during driving and until incorporation in the structure the free length of the pile shall be adequately supported and restrained by means of leaders, trestles, temporary supports or other guide arrangements to maintain position and alignment and to prevent buckling. In marine works, lengths which remain unsupported after driving shall be adequately restrained until incorporated into the Permanent Works. These constraint arrangements shall be such that no significant damage occurs to piles or their coatings.

### 6.11.2. Performance of driving equipment

The Contractor shall provide the Engineer with information on the efficiency and energy of the driving equipment including when followers are used. Where required in the Particular Specification, dynamic evaluation and analysis shall be provided.

Where a drop hammer is used, the mass of the hammer shall be at least half that of the pile. For other types of hammer the energy delivered to the pile per blow shall be at least equivalent to that of a drop hammer of the stated mass. Drop hammers shall not be used from floating craft in such a manner as to cause instability of the craft or significant damage to the pile.

### 6.11.3. Length of piles

The length of pile to be driven and any additional lengths of pile to be added during driving shall comply with the Contractor's schedule unless otherwise specified in the Particular Specification. During the execution of the Works any amendments to the scheduled lengths shall be reported to the Engineer giving adequate notice prior to installation of the piles concerned.

### 6.11.4. Driving procedure and redrive checks

The driving procedure shall be such as to avoid damage to the piles.

The Engineer shall be given 24 hours' notice of commencement of the driving of the first pile.

The Contractor shall give adequate notice and provide all necessary facilities to enable the Engineer to check driving resistances, sets and temporary compressions.

The driving of each pile shall be continuous until the depth or set as required by the design has been reached. In the event of unavoidable interruption to driving, the pile may be redriven provided it can be driven to the specified depth and/or resistance or set without damage. A follower shall not be used unless the set where applicable is revised in order to take into account reduction in the effectiveness of the hammer blow.

Driving records shall be made for every pile. This record shall contain the weight and fall of the hammer or ram and the number of blows for each 0.25 m of penetration, unless otherwise specified in the Particular Specification.

The Contractor shall inform the Engineer without delay if an unexpected change in driving characteristics is noted.

### 6.11.5. Set

The sets and temporary compression shall be measured and recorded for each pile at the completion of driving unless otherwise stated in the Particular Specification.

When a set or resistance is being measured, the following requirements shall be met.

(a) The exposed part of the pile shall be in good condition, without damage or distortion.

(b) The dolly and packing, if any, shall be in sound condition.

(c) The hammer blow shall be in line with the pile axis and the impact surfaces shall be flat and at right angles to the pile hammer axis.

(d) The hammer shall be in good condition, delivering adequate energy per blow and operating correctly.

(e) The temporary compression of the pile shall be recorded, if required in the Particular Specification.

The set shall be recorded either as the penetration in millimetres per 10 blows or as the number of blows required to produce a penetration of 25 mm.

### 6.11.6. Driving sequence and risen piles

Piles shall driven in a sequence to minimize any detrimental effects of heave and lateral displacement of the ground.

The sequence and method of piling including preboring shall limit uplift and lateral movement so that the final position of each pile is within the specified tolerances. At all times the deflections of each pile from its axis as formed shall not be such as to cause damage or impair durability of the piles or any structures or services.

The maximum permitted uplift of each pile due to any one pile driven within a pile centre to centre radius of eight pile diameters is 3 mm unless it can be demonstrated by static load testing that uplift exceeding this amount does not affect the ability of the pile to meet the requirements of this Specification.

Even if during the installation of preliminary piles uplift is shown to be within the permitted maximum and the preliminary piles tested meet the requirements of the Specification checks of uplift on working piles shall be made by the Contractor at least once a week throughout the period of the piling work and the results reported to the Engineer.

If preliminary piles are not installed the Contractor shall commence installation of working piles taking measures to reduce or eliminate uplift until it can be established by site measurements that such measures are no longer necessary. Thereafter checks on uplift shall be made by the Contractor at least once a week and the results reported to the Engineer.

If a static load test shows that a pile which is uplifted more than the maximum permitted amount due to any pile driven within a radius of eight pile diameters does not comply with the requirements of the Specification all such piles that may have been uplifted shall be redriven.

If records and measurements show that piles have been laterally displaced so as to be outside the permitted tolerance, damaged or are of impaired durability the measures the Contractor plans to adopt to enable the piles to comply with the Specification shall be provided to the Engineer.

Laterally displaced piles shall not be corrected by forceable correction at the heads, unless the Contractor can demonstrate that the integrity, durability and performance of the piles has not been adversely affected.

### 6.11.7. Preboring

If preboring is specified the pile shall be pitched after preboring to the depth and diameter stipulated in the Particular Specification.

Other means to ease pile drivability may be used provided the completed piles meet the requirements of the Specification.

## 6.12. Preparation of pile heads

If a steel structure is to be welded to piles, the piles shall be cut square and to within 5 mm of the levels specified. If pile heads are to be encased in concrete they shall be cut to within 20 mm of the levels specified, and protective coatings shall be removed from the surfaces of the pile heads down to a level 100 mm above the soffit of the concrete.

*Specification for piling and embedded retaining walls.* Thomas Telford, London, 1996.

43

# 7. Timber piles

## 7.1. General

All materials and work shall be in accordance with Sections 1 and 7 of this Specification, except where there may be conflict of requirements, in which case this Section shall take precedence.

## 7.2. Particular Specification

The following matters are, where appropriate, described in the Particular Specification:

    (*a*)   specified working load
    (*b*)   additional performance criteria for piles under tests (also see Table 1.3)
    (*c*)   species and grades of timber
    (*d*)   details of pile encasement, where required for shaft length above the water table
    (*e*)   length and dimensions
    (*f*)   preservative treatment
    (*g*)   surface finishes to piles
    (*h*)   grades and types of pile shoes
    (*i*)   splicing
    (*j*)   penetration or depth or toe level
    (*k*)   driving resistance or dynamic evaluation or set
    (*l*)   trial drives
    (*m*)   preliminary piles
    (*n*)   uplift/lateral displacement trials
    (*o*)   preboring or jetting or other means of easing pile driveability
    (*p*)   detailed requirements for driving records (including requirements for redrives)
    (*q*)   disposal/ownership of cut-off heads of piles
    (*r*)   other particular requirements.

## 7.3. Materials
### 7.3.1. Species of timber

Timber shall be as specified. If hardwood is to be used the Contractor shall provide documentation to the Engineer verifying that the timber has been taken from a managed or recognized source.

### 7.3.2. Grade of timber

The timber shall be new and free from defects, decay, large, loose or dead knots, undue shakes or excessive sap on more than one edge which may affect the strength or durability of piles. The grain shall be straight and parallel to the axis of the pile.

For timbers other than tropical hardwoods the grade as described in BS 4978 shall be Special Structural (SS) Grade. For tropical hardwoods the grade of the timber as described in BS 5756 shall be Hardwood Structural (HS) Grade.

### 7.3.3. Sapwood

Tree trunks for use as round piles shall have the bark removed but the sapwood left in place. They shall be treated with preservative where specified. Sawn or hewn softwood or hardwood which is to be treated with preservative need not have the sapwood removed. Hardwood which is to be used untreated shall be free of sapwood.

### 7.3.4. Tolerance in timber dimensions

The tolerance on cross-sectional dimensions of sawn piles in relation to the dimensions specified shall be −6 mm and +12 mm. The centroid of any cross-section of a sawn pile shall not deviate by more than 25 mm from the straight line connecting the centroids of the end faces of the pile.

Hewn piles may be tapered, in which case the taper, as determined by the difference in cross-sectional dimension between the ends of the pile divided by the length, shall not exceed 1 in 100. At mid-length of a hewn pile the cross-sectional dimension shall be within 20 mm of the specified dimension. The centroid of any cross-section of a hewn pile shall not deviate by more than 25 mm from a straight-line chord joining the centroids of pile sections at 6 m centres. The minimum cross-sectional dimension of a hewn pile shall be the mid-length specified dimension less 1/100 of the half-length of the pile without further reduction for tolerance.

### 7.3.5. Condition of timber

The timber shall be free from rot, pests, fungal or pest attack and from defects not permitted for its grade. Timber to be treated with preservative shall have a moisture content of not more than that stated in BS 913 or BS 4072. Timber not to be treated with preservative shall have a moisture content of not more than 23% at the time of installation, unless it is placed in a permanently wet position before fungal growth can begin.

### 7.3.6. Preservatives

Coal tar creosote shall comply with BS 144. Water-borne copper/chrome/arsenic compositions shall comply with BS 4072, type 1 or type 2 as specified.

### 7.3.7. Pile shoes

The material and dimensions of the pile shoes shall be as specified.

Cast iron shoes shall be made from chill-hardened iron as used for making grey iron castings to BS 1452, Grade 14, or ductile iron in accordance with BS 2789. The chilled iron point shall be free from major blow-holes and other surface defects.

Cast steel pile shoes shall be fabricated from steel to BS 3100, Grade A1.

Fabricated shoes and their fastenings shall be made from steel to BS 4360, Grade 43A or 50A.

Straps and other fastenings to cast pile shoes shall be of steel to BS 4360, Grade 43A, and shall be cast into the point to form an integral part of the shoe.

## 7.4. Inspection and stacking

The Contractor shall notify the Engineer of the delivery of timber piles to the Site or to the place of preservative treatment, and provide all labour and materials to enable the Engineer to inspect each piece on all faces and to measure it at the time of unloading and immediately prior to driving.

Timber shall be marked and stacked in lengths on paving or drained hard ground. Each piece of timber shall be clear of the ground and have an air space around it. The baulks or piles shall be separated by suitable blocks or spacers placed vertically one above the other and positioned at centres which are close enough to prevent sagging. The timber shall be protected from the sun and rain by means of roofing over with tarpaulins or other appropriate covering which allows free circulation of air.

## 7.5. Treatment with preservative

Preservative treatment shall be carried out in accordance with BS 913 or BS 4072 as specified. Cutting and boring of timber shall be done as far as possible before preservative treatment, but where this is impracticable all surfaces subsequently cut or bored shall be heavily coated with preservative as specified in the relevant British Standard for preservative treatment or in accordance with the manufacturer's instructions as appropriate.

Certificates of treatment must be obtained and presented to the Engineer for all treated timber. The type and method of treatment must be compatible with the type of timber and the use to which the timber so treated is to be put.

## 7.6. Pile shoes

The shoes shall be attached to the pile by steel straps fixed, spiked, screwed or bolted to the timber. The shoes shall be coaxial with the pile and firmly bedded to it.

## 7.7. Pile heads

The pile head shall be flat and at right angles to the axis of the pile.

Unless otherwise specified the head of each pile shall be trimmed to a round cross-section and fitted with a tight steel ring. The ring shall be not less than 50 mm deep by 12 mm thick in cross-section and the join shall be welded for its full section. The external diameter of the ring shall be that of at least allowable transverse dimension of the head of the pile. The top of the ring shall be between 10 mm and 20 mm from the top of the pile. If the ring is displaced during driving it shall be refitted. If the ring is broken a new ring shall be fitted.

As an alternative to a ring, a metal helmet may be used, the top of the pile being trimmed to fit closely into the recess of the underside of the helmet. A hardwood dolly and, if necessary, a packing shall be used above the helmet.

If during driving the head of the pile becomes excessively broomed or otherwise damaged, the damaged part shall be cut off, the head retrimmed and the ring or helmet refitted.

## 7.8. Splicing

Piles shall be provided in one piece unless otherwise specified. A splice shall be capable of resisting safely any stresses which may develop during lifting, pitching or driving, and under the design verification load. The position and details of the splice shall be as specified.

The splice shall be made as follows. The two timbers shall be of the same sectional dimensions and each cut at right angles to its axis to make contact over the whole of the cross-sectional when the two timbers are coaxial. A jointing compound shall be used at the contact surface. Round timbers shall be joined by a section of steel tube. Rectangular piles shall be joined by a prefabricated steel box section fitting the timbers closely or by steel splice plates. The connection shall be bolted, screwed or spiked to the timbers to keep the joined ends in close contact. The two parts shall not be more than 1:100 out of axial alignment.

Where it is necessary to extend a partly driven pile, the upper part must be securely supported during the making of the joint.

## 7.9. Driving piles
### 7.9.1. Leaders and trestles

At all stages during driving and until incorporation in the super-structure the pile shall be adequately supported and restrained by means of leaders, trestles, temporary support or other guide arrangements to maintain the position and alignment and to prevent bending. The arrangements shall be such that damage to the piles does not occur.

### 7.9.2. Performance of driving equipment

The Contractor shall provide the Engineer with information on the efficiency and energy of the driving equipment including when followers are used.

Where a drop hammer or other impact hammer is used, the mass of the falling hammer or ram shall be at least half that of the pile. Drop hammers shall not be used from floating craft in such a manner as to cause instability of the craft.

### 7.9.3. Length of piles

The length of pile supplied to be driven in any position and any additional lengths to be added during driving shall comply with the Specification. During the execution of the Works any changes to the supplied lengths shall be made known to the Engineer.

### 7.9.4. Driving procedure and redrive checks

The Contractor shall inform the Engineer without delay if an unexpected change in driving characteristics is noted.

Driving records shall be made for every pile. This record shall contain the weight and fall of the hammer or ram and the number of blows for each 0.25 m of penetration.

The Contractor shall give adequate notice and provide all necessary facilities to enable the Engineer to check driving resistances, sets and temporary compressions.

The Engineer shall be given 24 hours' notice of commencement of the driving of the first pile.

The driving procedure shall be such as to avoid damage to the piles.

### 7.9.5. Set

If required by the Particular Specification, sets and temporary compressions shall be measured for each pile at 0.25 m intervals from the time of marked increase in the driving resistance until each pile reaches its final level.

When a set or resistance is being measured, the following requirements shall be met.

(a) The exposed part of the pile shall be in good condition, without damage or distortion.

(b) The dolly and packing, if any, shall be in sound condition.

(c) The hammer blow shall be in line with the pile axis and the impact surfaces shall be flat and at right angles to the pile and hammer axis.

(d) The hammer shall be in good condition, delivering adequate energy per blow and operating correctly.

(e) The temporary compression of the pile shall be recorded.

### 7.9.6. Spliced piles

Spliced piles shall be observed continuously during driving to detect any departure form true alignment of the two parts. If any such departure occurs, driving shall be suspended and the Engineer shall be informed.

### 7.9.7. Driving sequence and risen piles

Piles shall be driven in a sequence to minimize the detrimental effects of heave and lateral displacement of the ground.

The sequence and method of piling including preboring shall limit uplift and lateral movement so that the final position of each pile is within the specified tolerances. At all times the deflections of each pile from its axis as formed shall not be such as to cause damage or impair durability of the piles or any structures or services.

The maximum permitted uplift of each pile due to any one pile driven within a pile centre to centre radius of eight pile diameters is 3 mm unless it can be demonstrated that uplift exceeding this amount does not affect the ability of the pile to meet the requirements of this Specification.

Even if during the installation of preliminary piles uplift is shown to be within the permitted maximum and the preliminary piles tested meet the requirements of the Specification periodic checks of uplift on working piles shall be made by the Contractor throughout the period of the piling work and the results reported to the Engineer.

If preliminary piles are not installed the Contractor shall commence installation of working piles taking measures to reduce or eliminate uplift until it can be established by site measurements that such measures are no longer necessary. Thereafter periodic checks on uplift shall be made by the Contractor and the results reported to the Engineer.

If a static load test shows that a pile which is uplifted more than the maximum permitted amount due to any pile driven within a radius of eight pile diameters does not comply with the requirements of the Specification all such piles that may have been uplifted shall be redriven.

If records and measurements show that piles have been laterally displaced so as to be outside the permitted tolerance, damaged or are of impaired durability the measures the Contractor proposes to adopt to enable the piles to comply with the Specification shall be provided to the Engineer.

Laterally displaced piles shall not be corrected by jacking at the heads.

### 7.9.8. Preboring

If preboring is specified the pile shall be pitched after preboring to the depth and diameter shown on the Drawings or stipulated in the Particular Specification.

## 7.10. Preparation of pile heads

After driving, the piles shall be cut off square to sound timber to within 5 mm of the levels shown on the drawings and the cut surfaces shall be heavily coated with preservative if and as specified for the initial treatment.

*Specification for piling and embedded retaining walls.* Thomas Telford, London, 1996.

# 8. Reduction of friction on piles

**8.1. General**

Where friction reduction is specified but the particular method of reducing friction is not specified, the Contractor shall provide full details of the method which he plans to employ. The Contractor shall ensure that any product used will be compatible with the ground conditions into which it will be installed. Particular requirements are detailed in the Particular Specification.

**8.2. Particular Specification**

The following matters are, where appropriate, described in the Particular Specification:

- (a) the type and particular description of method to be used
- (b) the numbers or other identification of piles to be treated to reduce friction
- (c) the length of pile to be treated
- (d) preparatory preboring or other work necessary for proper application of the method
- (e) depth, diameter and means of ensuring temporary stability of any preboring where required
- (f) designated manufacturer's name and details where a proprietary product is required
- (g) testing piles or trial piles to demonstrate the effectiveness of the method
- (h) other particular requirements.

**8.3. Pre-applied bituminous or other friction-reducing coating materials**

*8.3.1. General*

Where a proprietary product is used, the process of cleaning pile surfaces, and the conditions and methods of application shall conform with the manufacturer's current instructions. All materials shall conform with the manufacturer's specification, which shall be given to the Engineer before any coating is applied.

*8.3.2. Protection from damage*

Where a friction-reducing material has been applied to a pre-formed pile prior to installation, it shall be protected from damage during handling and transportation. In the event of damage to the coating, it shall be made good on site to the same specification as the original coating prior to the pile being driven. Where bituminous materials are involved, precautions shall be taken as necessary in hot weather to prevent excessive flow or displacement of the coating. The coated piles shall be adequately protected against direct sunlight and, if stacked, they shall be separated to prevent their coatings sticking together.

*8.3.3. Pile driving*

In the case of applied coatings, the piles shall not be driven when the air temperature is such that the coating will crack, flake or otherwise be damaged prior to entry into the ground. Where bituminous materials are involved, driving shall be carried out while the temperature is at or above 5°C or as called for in the manufacturer's instructions.

**8.4. Pre-applied low-friction sleeving**

Detailed design of the pre-applied low-friction sleeving shall be submitted to the Engineer prior to driving the piles. All materials

---

shall conform with the manufacturer's specification, which shall be given to the Engineer before the material is used.

### 8.5. Formed-in-place low-friction surround

Detailed design of the formed-in-place low-friction surround shall be submitted to the Engineer prior to driving or forming the piles. All materials shall conform with the manufacturer's specification, which shall be given to the Engineer before the material is used.

### 8.6. Pre-installed low-friction sleeving

Detailed design of the pre-installed low-friction sleeving shall be submitted to the Engineer prior to driving or forming the piles. All materials shall conform with the manufacturer's specification, which shall be given to the Engineer before the material is used.

### 8.7. Inspection

The Engineer may call for piles to be partially exposed or extracted. Where significant damage to the coating is found to have occurred the Contractor shall submit a method statement for the repair or replacement of the coating.

# 9. Non-destructive methods for testing piles

## 9.1. Integrity testing of piles
### 9.1.1. Method of testing

Where integrity testing is called for, the method to be adopted shall be one of the following, as specified:

(a) impulse method
(b) Sonic Echo, Frequency Response or Transient Dynamic steady state vibration method
(c) sonic logging method.

Other methods may be considered by the Engineer subject to satisfactory evidence of performance.

### 9.1.2. Particular Specification

The following matters are, where appropriate, described in the Particular Specification:

(a) the method of test to be carried out
(b) the number, type and location of piles to be tested
(c) the stages in the programme of works when a phase of integrity testing is to be carried out
(d) the number and location of piles in which ducts are to be placed and number and length of ducts to be provided in each pile for the sonic logging method
(e) the time after testing at which the test results and findings shall be available to the Engineer, if different from the requirements of Clause 9.1.7
(f) where sonic coring is called for, the depth of pile over which the testing is required, the depth intervals to be not greater than 0.25 m
(g) the number of days to elapse between pile casting and integrity testing
(h) preparation of pile head for testing using the vibration method
(i) other particular requirements.

### 9.1.3. Age of piles at time of testing

In the case of cast-in-place concrete piles, integrity tests shall not be carried out until the number of days specified in the Particular Specification have elapsed since pile casting.

### 9.1.4. Preparation of pile heads

Where the method of testing requires the positioning of sensing equipment on the pile head, the head shall be broken down to expose sound concrete and shall be clean, free from water, laitance, loose concrete, overspilled concrete and blinding concrete and shall be readily accessible for the purpose of testing.

### 9.1.5. Specialist Sub-contractor

The testing shall be carried out by a specialist firm, subject to demonstration to the Engineer of satisfactory performance on other similar contracts before the commencement of testing.

The Contractor shall submit to the Engineer the name of the specialist integrity testing firm, a description of the test equipment, a test method statement and a programme for executing the specified tests prior to commencement of the Works.

### 9.1.6. Interpretation of tests

The interpretation of tests shall be carried out by competent and experienced persons.

The Contractor shall give all available details of the ground

*Specification for piling and embedded retaining walls.* Thomas Telford, London, 1996.

51

conditions, pile dimensions and construction method to the specialist firm before the commencement of integrity testing in order to facilitate interpretation of the tests.

### 9.1.7. Report

Preliminary results of the tests shall be submitted to the Engineer within 24 hours of carrying out the tests.

The test results and findings shall be reported to the Engineer within 10 days of the completion of each phase of testing.

The report shall contain a summary of the method of interpretation including all assumptions, calibrations, corrections, algorithms and derivations used in the analyses. If the results are presented in a graphical form, the same scales shall be used consistently throughout the report. The units on all scales shall be clearly marked.

### 9.1.8. Anomalous results

In the event that any anomaly in the acoustic signal is found in the results indicating a possible defect in the pile the Contractor shall report such anomalies to the Engineer immediately. The Contractor shall demonstrate to the Engineer that the pile is satisfactory for its intended use or shall carry out remedial works to make it so. Sonic logging tubes shall be grouted up after the Contractor has demonstrated that the pile is satisfactory.

### 9.2. Dynamic pile-testing
### 9.2.1. Particular Specification

The following matters are, where appropriate, described in the Particular Specification:

(a) the number, type and location of piles to be tested
(b) the stages in the programme of works when a phase of dynamic testing is to be carried out
(c) the minimum dynamic test load
(d) the time period following installation at which testing is required
(e) measurement of set and temporary compression
(f) details of work to be carried out on a pile head following a test.

### 9.2.2. Measuring instruments

Current calibration certificates shall be provided to the Engineer for all instruments and monitoring equipment before testing commences.

### 9.2.3. Hammer

The hammer and all other equipment used shall be capable of delivering an impact force sufficient to mobilize the equivalent specified dynamic test load without damaging the pile.

### 9.2.4. Preparation of the pile head

The preparation of the pile head for the application of the dynamic test load shall involve trimming the head, cleaning and building up the pile using materials which will at the time of testing safely withstand the impact stresses. The impact surface shall be flat and at right angles to the pile axis. Where pile preparation requires drilling holes or welding, this preparation shall not adversely affect the performance of the pile when in service.

### 9.2.5. Interpretation of tests

The interpretation of the tests shall be carried out by competent and experienced persons.

The Contractor shall give all available details of the ground conditions, pile dimensions and construction method to the specialist firm carrying out the testing in order to facilitate interpretation of tests.

### 9.2.6. Time of testing

The time between the completion of installation and testing for a preformed pile shall be more than 12 hours, and in the case of a cast-in-place pile shall be such that the pile is not damaged under the impact stresses.

### 9.2.7. Measurement of set

If specified, the permanent penetration per blow and temporary compression of the pile and soil system shall be measured independently of the instruments being used to record the dynamic test data from a fixed reference point unaffected by the piling operations.

### 9.2.8. Results

Initial results shall be provided to the Engineer within 24 hours of the completion of a test. These shall include:

(a) the maximum force applied to the pile head
(b) the maximum pile head velocity
(c) the maximum energy imparted to the pile.

Subsequently, a full report shall be given to the Engineer, within 10 days of the completion of testing, including:

(a) date of pile installation
(b) date of test
(c) pile identification number and location
(d) length of pile below commencing surface
(e) total pile length, including projection above commencing surface at time of test
(f) length of pile from instrumentation position to toe
(g) hammer type, drop and other relevant details
(h) blow selected for analysis
(i) test load achieved (i.e. total mobilized deduced static load)
(j) pile head movement at equivalent Design Verification Load
(k) pile head movement at equivalent Design Verification Load plus 50% of Specified Working Load
(l) pile head movement at maximum applied test load
(m) permanent residual movement of pile head after each blow
(n) temporary compression.

For all piles tested, the following information shall be provided for typical blows:

(a) date of pile installation
(b) date of test
(c) pile identification number and location
(d) length of pile below commencing surface
(e) total pile length, including projection above commencing surface at time of test
(f) length of pile from instrumentation position to toe
(g) hammer type, drop and other relevant details
(h) permanent set per blow
(i) maximum force at pile head
(j) maximum velocity at pile head
(k) maximum downward energy imparted to the pile
(l) dynamic soil resistance mobilized during the blow
(m) static soil resistance mobilized during the blow assuming that soil damping is proportional to pile velocity
(n) magnitude and location of possible pile damage.

For piles selected by the Engineer, an analysis of measurements from selected blows shall be carried out using a numerical model of the pile and soil to provide the following information:

(*a*)  magnitude and distribution of mobilized static soil resistance

(*b*)  magnitude and distribution of soil stiffness and damping

(*c*)  deduced static load deflection behaviour of the pile at the head and toe

(*d*)  assumptions made in the analysis

(*e*)  limitations of the method.

# 10. Static load testing of piles

**10.1. General**

The design and construction of the load application system shall be satisfactory for the required test. These details shall be submitted to the Engineer prior to the commencement of testing.

**10.2. Particular Specification**

The following matters are, where appropriate, described in the Particular Specification:

(a) type of pile
(b) type of test
(c) loads to be applied and procedure to be adopted in testing preliminary piles
(d) loads to be applied in proof-testing of working piles
(e) special materials to be used in construction of preliminary test piles where appropriate
(f) special construction detail requirements for test piles
(g) special requirements for pile-testing equipment and arrangement
(h) pile installation criteria
(i) time interval between pile installation and testing
(j) removal of Temporary Works
(k) special requirements for the application of a lateral load to a pile detailed in accordance with the expected conditions of loading
(l) details of work to be carried out to the test pile cap or head at the completion of a test
(m) other particular requirements.

**10.3. Construction of a pile to be tested**

**10.3.1. Notice of construction**

The Contractor shall give the Engineer at least 48 hours' notice of the commencement of construction of any preliminary pile which is to be test-loaded.

**10.3.2. Method of construction**

Each preliminary test pile shall be constructed in a manner similar to that to be used for the construction of the working piles, and by the use of similar equipment and materials. Extra reinforcement and concrete of increased strength will be permitted in the shafts of preliminary piles provided prior notification is made.

**10.3.3. Boring or driving record**

For each preliminary pile which is to be tested a detailed record of the conditions experienced during boring, or of the progress during driving, shall be made and submitted daily, not later than noon on the next working day. Where the Engineer requires soil samples to be taken or in-situ tests to be made, the Contractor shall present the results without delay.

**10.3.4. Concrete test cubes**

Four test cubes shall be made from the concrete used in the preliminary test pile and from the concrete used for building up a working pile. If a concrete pile is extended or capped for the purpose of testing, a further four cubes shall be made from the corresponding batch of concrete. The cubes shall be made and tested in accordance with BS 1881.

The pile test shall not be started until the strength of the cubes taken from the pile exceeds twice the average direct stress in any pile section under the maximum required test load, and the strength of the cubes taken from the cap exceeds twice the average stress at any point in the cap under the same load.

### 10.3.5. Preparation of working pile to be tested

If a test is required on a working pile the Contractor shall cut down or otherwise prepare the pile for testing as required by the Engineer in accordance with the Particular Specification and Clauses 10.5.5 and 10.5.6.

### 10.3.6. Cut-off level

The cut-off level for a preliminary test pile shall be as specified in the Particular Specification.

Where the cut-off level of working piles is below the ground level at the time of pile installation and where it is required to carry out a proof test from that installation level, either allowance shall be made in the determination of the design verification load for friction which may be developed between the cut-off level and the existing ground level, or the pile may be sleeved appropriately or otherwise protected to eliminate friction which develops over the extended length.

## 10.4. Supervision

The setting-up of pile testing equipment shall be carried out under competent supervision and the equipment shall be checked to ensure that the setting-up is satisfactory before the commencement of load application.

All tests shall be carried out only under the direction of an experienced and competent supervisor conversant with the test equipment and test procedure. All personnel operating the test equipment shall have been trained in its use.

## 10.5. Safety precautions
### 10.5.1. General

Design, erection and dismantling of the pile test reaction system and the application of load shall be carried out according to the requirements of the various applicable statutory regulations concerned with lifting and handling heavy equipment and shall safeguard operatives and others who may from time to time be in the vicinity of a test from all avoidable hazards.

### 10.5.2. Kentledge

Where kentledge is used the Contractor shall construct the foundations for the kentledge and any cribwork, beams or other supporting structure in such a manner that there will not be differential settlement, bending or deflexion of an amount that constitutes a hazard to safety or impairs the efficiency of the operation. The kentledge shall be adequately bonded, tied or otherwise held together to prevent it from becoming unstable because of deflexion of the supports or for any other reason.

The weight of kentledge for each test shall be greater than the maximum test load for that test, and if the weight is estimated from the density and volume of the constituent materials an adequate factor of safety against error shall be allowed. Additional kentledge required shall be determined taking into account the accuracy of positioning of the centre of gravity of the stack.

### 10.5.3. Tension piles, reaction piles and ground anchorages

Where tension piles, reaction piles or ground anchorages are used to provide the necessary load reaction, they shall be so designed that they will resist the forces applied to them safely and without excessive deformation which could cause a safety hazard during the work. Such piles or anchorages shall be placed in the specified positions, and bars, tendons or links shall be aligned to give a stable reaction in the direction required. Any welding employed to extend or to fix anchorages to a reaction frame shall be carried out so that the full strength of the system is adequate and unimpaired.

### 10.5.4. Testing equipment

In all cases the Contractor shall ensure that when the hydraulic jack and load-measuring device are mounted on the pile head the whole system will be stable up to the maximum load to be applied.

If in the course of carrying out a test any unforeseen occurrence should take place, further loading shall not be applied until a proper engineering assessment of the condition has been made and steps have been taken to rectify any fault. Reading of gauges should, however, be continued where possible and if it is safe to do so.

Where an inadequacy in any part of the system might constitute a hazard, means shall be provided to enable the test to be controlled from a position clear of the kentledge stack or test frame.

The hydraulic jack, pump, hoses, pipes, couplings and other apparatus to be operated under hydraulic pressure shall be capable of withstanding a pressure of $1\frac{1}{2}$ times the maximum pressure used in the test without leaking.

The maximum test load expressed as a reading on the gauge in use shall be displayed and all operators shall be made aware of this limit.

### 10.5.5. Pile head for compression test

For a pile that is tested in compression, the pile head or cap shall be formed to give a plane surface which is normal to the axis of the pile, sufficiently large to accommodate the loading and settlement measuring equipment and adequately reinforced or protected to prevent damage from the concentrated application of load from the loading equipment.

Any pile cap shall be concentric with the test pile; the joint between the cap and the pile shall have a strength equivalent to that of the pile.

Sufficient clear space shall be made under and around any part of the cap projecting beyond the section of the pile so that, at the maximum expected settlement, load is not transmitted to the ground by the cap.

### 10.5.6. Pile connection for tension test

For a pile that is tested in tension, means shall be provided for transmitting the test load axially without inducing moments in the pile. The connection between the pile and the loading equipment shall be constructed in such a manner as to provide a strength equal to the maximum load which is to be applied to the pile during the test, with an appropriate factor of safety on the structural design.

### 10.6. Reaction systems
### 10.6.1. Compression tests

Compression tests shall be carried out using kentledge, tension piles or specially constructed anchorages. Kentledge shall not be used for tests on raking piles.

Where kentledge is to be used, it shall be supported on cribwork and positioned so that the centre of gravity of the load is as close as possible to the axis of the pile. The bearing pressure under supporting cribs shall be such as to ensure stability of the kentledge stack.

### 10.6.2. Tension tests

Tension tests may be carried out using compression piles, rafts or grillages constructed on the ground to provide the necessary reaction. In all cases the resultant force of the reaction system shall be coaxial with the test pile.

Where inclined piles or reactions are specified or their use is planned by the Contractor, full details shall be submitted prior to the commencement of testing.

### 10.6.3. Working piles

If the Contractor plans to use working piles as reaction piles, he shall notify the Engineer of his intention prior to commencement of work. Working reaction piles shall not uplift by more than half their specified permissible settlement at working load. The integrity of all working piles used as reaction piles shall be checked on completion of static load testing.

Where working piles are used as reaction piles their movement shall be measured and recorded to within an accuracy of 0.5 mm.

### 10.6.4. Spacing

Where kentledge is used for loading vertical piles in compression, the distance from the edge of the test pile to the nearest part of the crib supporting the kentledge stack in contact with the ground shall be not less than 1.3 m.

The centre-to-centre spacing of vertical reaction piles, including working piles used as reaction piles, from a test pile shall be not less than three times the diameter of the test pile or the reaction piles or 2 m whichever is the greatest. Where a pile to be tested has an enlarged base, the same criterion shall apply with regard to the pile shaft, with the additional requirement that no surface of a reaction pile shall be closer to the base of the test pile than one half of the enlarged base diameter. Where vertical reaction piles penetrate deeper than the test pile, the centre-to-centre spacing of the reaction piles from the test pile shall be not less than five times the diameter of the test pile or the reaction piles whichever is the greatest unless the base capacity of the test pile is less than 20% of the total ultimate capacity.

Where ground anchorages are used to provide a test reaction for loading in compression, no section of fixed anchor length transferring load to the ground shall be closer to the test pile than three times the diameter of the test pile. Where the pile to be tested has an enlarged base the same criterion shall apply with regard to the pile shaft, with the additional requirement that no section of the fixed anchor transferring load to the ground shall be closer to the pile base than a distance equal to the base diameter.

### 10.6.5. Adequate reaction

The reaction frame support system shall be adequate to transmit the maximum test load in a safe manner without excessive movement or influence on the test pile. Calculations shall be provided to the Engineer when required to justify the design of the reaction system.

### 10.6.6. Care of piles

The method employed in the installation of the reaction system shall be such as to prevent damage to any test pile or working pile.

## 10.7. Equipment for applying load

The equipment used for applying load shall consist of a hydraulic ram or jack. The jacking system shall be arranged in conjunction with the reaction system to deliver an axial load to the test pile and maintain it constant when required. The complete system shall be capable of safely transferring the maximum load required for the test. The length of stroke of a ram shall be sufficient to cater for deflection of the reaction system under load plus a deflection of the pile head up to 15% of the pile shaft diameter unless otherwise specified in the Particular Specification.

## 10.8. Measurement of load

The test load shall be measured by a single load cell or proving ring calibrated in divisions not exceeding 1% of the maximum load to be applied. If an electronic transducer is used each reading shall be immediately stored magnetically so that in case of power failure the readings are not lost.

The load cell or proving ring shall be calibrated immediately prior to the test and a certificate of calibration shall be submitted to the Engineer.

All increments of load shall be maintained to within 1% of the specified load.

A spherical seating of appropriate size shall be used to avoid eccentric loading. Care shall be taken to avoid any risk of buckling of the load application and measuring system. Load measuring and application devices shall be short in axial length in order to secure stability. The Contractor shall ensure that axial loading is maintained.

## 10.9. Control of loading

The loading equipment shall enable the load to be increased or decreased smoothly or to be held constant at any required value.

## 10.10. Measuring pile head movement
### 10.10.1. Maintained load test

In a maintained load test, movement of the pile head shall be measured by the method in Clause 10.10.3 and by one of the methods in Clause 10.10.4.

### 10.10.2. CRP and CRU test

In a CRP or a CRU test, movement of the pile head shall be measured by one of the methods in Clause 10.10.4.

### 10.10.3. Optical levelling method

An optical levelling method by reference to a remote datum may be used.

Where a level and staff are used, the level and scale of the staff shall be chosen to enable readings to be made to within an accuracy of 0.1 mm. A scale attached to the pile or pile cap may be used instead of a levelling staff. At least two reliable independent datum points shall be established. Each datum point shall be so situated as to permit a single setting-up position of the level for all readings.

No datum point shall be located where it can be affected by the test loading or other operations on the Site.

### 10.10.4. Reference beams and displacement gauges

An independent reference beam or beams shall be set up to enable measurement of the movement of the pile to be made to the required accuracy. The supports for a beam shall be founded in such a manner and at such a distance from the test pile and reaction system that movements of the ground do not cause movement of the reference beam or beams which will affect the accuracy of the test. The supports of the beam or beams shall be at least three test pile diameters or 2 m from the centre of the test pile, whichever distance is the greater. The beam must be free to move horizontally at one end.

Check observations of any movements of the reference beam or beams shall be made and a check shall be made of the movement of the pile head relative to a remote reference datum at the start and end and at maximum load for each loading cycle.

The measurement of pile movement shall be made by four dial gauges rigidly mounted on the reference beam or beams, bearing on prepared flat surfaces fixed to the pile cap or head and normal to the pile axis. Alternatively, the gauges may be fixed to the pile and bear on prepared surfaces on the reference beam or beams. The dial gauges shall be placed equidistant from the pile axis and from each other. The dial gauges shall enable readings to be made to an accuracy of at least 0.01 mm and have a stem travel of at least 25 mm. Machined spacer blocks may be used to extend the range of reading. Equivalent electrical displacement-measuring devices may be substituted.

The Contractor may submit details of any other method of measuring the movement of the test pile head. The method shall be accurate to within 0.01 mm of the pile head movement.

## 10.11. Protection of testing equipment
### 10.11.1. Protection from weather

Throughout the test period all equipment for measuring load and movement and beams shall be protected from adverse effects of sun, wind and precipitation. Temperature reading shall be taken at the start, end and at the maximum load of each loading cycle.

### 10.11.2. Prevention of disturbance

Construction activity and persons who are not involved in the testing process shall be kept at a sufficient distance from the test to avoid disturbance to the measuring apparatus. Full records shall be kept of any unavoidable activity and its effects.

## 10.12. Notice of test

The Contractor shall give the Engineer at least 24 hours' notice of the commencement of the test. No load shall be applied to the test pile before the commencement of the specified test procedure.

## 10.13. Test procedure
### 10.13.1. Proof load test procedure (working compression piles)

The maximum load which shall be applied in a proof test shall normally be the sum of the design verification load (DVL) plus 50% of the specified working load (SWL) The loading and unloading shall be carried out in stages as shown in Table 10.1. Any particular requirements given in the Particular Specification shall be complied with.

Following each application of an increment of load, the load shall be maintained at the specified value for not less than the period shown in Table 10.1 and until one of the following rate of settlement criteria is satisfied:

(i) the rate of settlement in a period of 30 minutes is less than 0.5% of the current cumulative settlement which has occurred

(ii) the rate of settlement is 0.05 mm or less in 30 minutes, whichever is the higher.

The rate of settlement shall be calculated from the slope of the line obtained by plotting values of settlement versus time and drawing a smooth curve through the points.

*Table 10.1. Minimum loading times for pile test*

| Load* | Minimum time of holding load |
| --- | --- |
| 25% DVL | 30 minutes |
| 50% DVL | 30 minutes |
| 75% DVL | 30 minutes |
| 100% DVL | 6 hours |
| 75% DVL | 10 minutes |
| 50% DVL | 10 minutes |
| 25% DVL | 10 minutes |
| 0 | 1 hour |
| 100% DVL | 1 hour |
| 100% DVL + 25% SWL | 1 hour |
| 100% DVL + 50% SWL | 6 hours |
| 100% DVL + 25% SWL | 10 minutes |
| 100% DVL | 10 minutes |
| 75% DVL | 10 minutes |
| 50% DVL | 10 minutes |
| 25% DVL | 10 minutes |
| 0 | 1 hour |

*SWL denotes specified working load; DVL denotes design verification load.

For any period when the load is constant, time and settlement shall be recorded immediately on reaching the load, at not more than 5 minute intervals up to 15 minutes, at approximately 15 minutes intervals up to 1 hour, at 30 minute intervals between 1 hour and 4 hours, and 1 hour intervals between 4 hours and 12 hours after the application of the increment of load.

Where the methods of measuring pile head movement given in Clauses 10.10.3 and 10.10.5 are used, the periods of time for which loads must be held constant to achieve the specified rates of settlement shall be extended as necessary to take into account the lower levels of accuracy available from these methods and to allow correct assessment of the settlement rate.

## 10.13.2. Test procedure for preliminary compression piles

The procedure to be adopted for carrying out preliminary load tests on compression piles shall be in accordance with the requirements specified in the Particular Specification and either the extended proof load test procedure or the constant rate of penetration testing procedure given below. A normal proof load test will constitute the first stage of such a preliminary test unless otherwise specified.

*Extended proof load test procedure*
Where verification of the required minimum load factor is called for in the Particular Specification, the loading procedure may be carried out as a continuation of the proof load testing procedure given in Clause 10.13.1.

Following the completion of the proof load test, the load shall be restored in two stages (DVL, DVL + 50% SWL), and shall subsequently be increased by stages of 25% of the specified working load or other specified amount until the maximum specified load for the test is reached. Following each application of an increment of load, the load shall be maintained at the specified value until one of the following rate of settlement criteria is satisfied:

(*i*)  the rate of settlement in a period of 30 minutes is less than 0.5% of the current cumulative settlement which has occurred, but not exceeding 0.12 mm in 30 minutes

(*ii*)  the rate of settlement is 0.05 mm or less in 30 minutes whichever is the higher.

The rate of settlement shall be calculated from the slope of the line obtained by plotting values of settlement versus time and drawing a smooth curve through the points.

The load shall then be reduced in five approximately equal stages to zero load, with penetration and load at each stage and at zero load being recorded.

*Constant rate of penetration (CRP) testing procedure*
The rate of movement of the pile head shall be maintained constant in so far as is practicable and shall be approximately 0.01 mm/s for piles in predominantly cohesive soils and 0.02 mm/s for piles in predominantly cohesionless soils.

Readings of loads, penetration and time shall be made simultaneously at regular intervals; the interval chosen shall be such that a curve of load versus penetration can be plotted without ambiguity.

Loading shall be continued until one of the following results is obtained

(*a*)  the maximum required test load as specified in the Particular Specification is reached *(b)* a constant or reducing load has been recorded for an interval of penetration of 10 mm

*(c)* a total movement of the pile head equal to 15% of the base diameter, or any other greater value of movement specified in the Particular Specification has been reached.

The load shall then be reduced in five approximately equal stages to zero load, with penetration and load at each stage and at zero load being recorded.

### 10.13.3. Testing of piles designed to carry load in tension

The testing of piles designed to carry load in tension shall follow the same procedure as specified in clauses 10.13.1 and 10.13.2.

The rate of movement of the pile head shall be maintained at approximately 0.005 mm/s in so far as is practicable.

## 10.14. Presentation of results
### 10.14.1. Results to be submitted

During the progress of a test, all records taken shall be available for inspection by the Engineer.

Results shall be submitted as follows:

*(a)* a preliminary copy of the test records to the Engineer, unless otherwise directed, within 24 hours of the completion of the test, which shall show

  *(i)* for a test by maintained load: for each stage of loading, the period for which the load was held, the load and the maximum pile movement at the end of the stage
  *(ii)* for a CRP or CRU test: the maximum load reached and a graph of load against penetration or load against uplift

*(b)* the completed schedule of recorded data as prescribed in clause 10.14.2 within 10 days of the completion of the test.

### 10.14.2. Schedule of recorded data

The Contractor shall provide information about the test pile in accordance with the following schedule where applicable.

*(a)* *General*
   Site location
   Contract identification
   Proposed structure
   Main contractor
   Piling contractor
   Engineer
   Client/Employer
   Date and time of test
*(b)* *Pile details*
   *All types of pile*
   Identification (number and location)
   Specified working load (SWL)
   Design verification load (DVL)
   Commencing surface level at pile position
   Head level at which test load was applied
   Type of pile
   Vertical or raking, compression or tension
   Shape and size of cross-section of pile, and position of any change in cross-section
   Shoe or base details
   Head details
   Length in ground

Level of toe
Dimensions of any permanent casing

*Concrete Piles*
Concrete mix/grade
Aggregate type and source
Cement type and cement replacement and type where used
Admixtures
Slump
Cube test results for pile and cap
Date of casting of precast pile
Reinforcement

*Steel piles*
Steel quality
Coating
Filling or core materials — type and quality

(c)  *Installation details*
*All piles*
Dates and times of boring, driving and concreting of test pile
Difficulties and delays encountered
Date and time of casting concrete pile cap

*Bored piles*
Type of equipment used and method of boring
Temporary casing — diameter, type and length
Full log of pile borehole
Method of placing concrete
Volume of concrete placed

*Driven preformed and driven cast-in-place piles*
Method of support of hammer and pile
Driven length of pile or temporary casing at final set
Hammer type, and size or weight
Dolly and packing, type and condition
Driving log (depth, hammer drop, blows per 250 mm, interruptions or breaks in driving)
Final set in number of blows to produce penetration of 25 mm
Redrive check, time interval and set in number of blows to produce penetration of 25 mm
At final set and at redrive set: for a drophammer or for a single acting hammer the length of the drop or stroke; for a diesel hammer the length of the stroke and the blows per minute; for a double acting hammer the operating pressure and the number of blows per minute.
Condition of pile head or temporary casing after driving
Use of a follower
Use of preboring
Use of jetting
Lengthening
Method of placing concrete

(d)  *Test procedure*
Mass of kentledge
Tension pile, ground anchorage or compression pile details
Plan of test arrangement showing position and distances of kentledge supports, rafts, tension or compression pile or

ground anchorages, and supports to pile movement reference system
Jack capacity
Method of load measurement
Method(s) of penetration or uplift measurement
Calibration certificates
Temperature readings

(e) *Test results*
In tabular form
In graphical form: load plotted against pile head movement, load plotted against time
Ambient temperature records during test

## 10.15. Completion of a test
### 10.15.1. Removal of test equipment

On completion of a test all measuring equipment and load application devices shall be dismantled and checked. All other test equipment, including kentledge, beams and supporting structures shall be removed from the test pile location. Measuring and other demountable equipment shall be stored in a safe manner so that it is available for further tests, if required, or removed from site.

Temporary tension piles and ground anchorages shall be cut off below ground level, and off-cut materials removed from the Site. The ground shall be made good to the original commencing surface level.

### 10.15.2. Preliminary test pile cap

Unless otherwise specified, the head of each preliminary test pile shall be cut off below ground level, off-cut material shall be removed from the Site and the ground made good to the original commencing surface level.

### 10.15.3. Proof test pile cap

On completion of a test on a proof pile, the test pile head shall be prepared as specified and left in a state ready for incorporation into the Permanent Works.

# 11. General requirements for embedded retaining walls

## 11.1. Definitions and Standards
### 11.1.1. Definitions

In this Series the terms 'submitted', 'demonstrated', 'notified' and 'required' mean 'submitted to the Engineer', 'demonstrated to the Engineer', 'notified to the Engineer' and 'required by the Engineer' respectively.

**Commencing surface:** is the level at which the piling equipment first enters the ground.

**Cut-off-level:** is the level to which the pile is trimmed.

**Element:** means an individual component utilized in a particular embedded retaining walling system, e.g. diaphragm wall panel, or primary or secondary pile in a secant wall, which can be constructed in isolation.

**Embedded retaining wall:** means retaining wall with shuttering provided either by the surrounding ground (i.e. cast against soil) or sheet piles inserted into the ground.

### 11.1.2. British Standards and other Codes of Practice

All materials and workmanship shall be in accordance with the appropriate British Standards, Codes of Practice and other specified standards current at the date of tender except where the requirements of these Standards or Codes of Practice are in conflict with this Specification in which case the requirements of this Specification shall take precedence.

## 11.2. Particular Specification

The following matters are, where appropriate, described in the Particular Specification:

  (*a*)  supervising officer
  (*b*)  location and description of the site
  (*c*)  nature of the works
  (*d*)  working area
  (*e*)  sequence of the works
  (*f*)  other works proceeding at the same time
  (*g*)  contract drawings
  (*h*)  office and other facilities for the Engineer
  (*i*)  submission of information (see Table 11.1)
  (*j*)  special requirements
  (*k*)  classes of loads on walls and excavation depths
  (*l*)  water exclusion requirements
  (*m*)  permissible limits on ground movements during wall installation and on wall movements during excavation and construction
  (*n*)  responsibility for design
  (*o*)  permissible damage criteria for existing critical structures or services.
  (*p*)  site datum and site grid
  (*q*)  restrictions on permissible working hours
  (*r*)  details of ground investigation reports
  (*s*)  additional ground investigation

*(t)* performance criteria for wall elements under test

*(u)* commencing surface level

*(v)* other particular requirements.

## 11.3. Progress Report

The Contractor shall submit to the Engineer on the first day of each week a progress report showing the current rate of progress and progress during the previous period on all important items of each section of the Works.

## 11.4. Wall layout, design and construction

*Option 1 — Contractor design*

The Contractor is required to design and construct the specified embedded retaining wall having the qualities of materials and workmanship specified and which meet the requirements of the Specification. The Contractor's design shall comprise the calculation of individual element sizes based on the ground conditions revealed by the site investigation.

Design details submitted at tender stage by the Contractor shall be as stipulated in the Particular Specification.

*Option 2 — Engineer design*

The Contractor is required to construct elements of the type(s) and dimensions specified and having the qualities of materials and workmanship specified.

## 11.5. Materials

The sources of supply of materials shall not be changed until the Contractor has demonstrated that the materials from the new source can meet all the requirements of the Specification.

Materials failing to comply with the Specification shall be removed promptly from the site.

## 11.6. Safety
### 11.6.1. Standards

Safety precautions through the piling operations shall comply with the Health and Safety at Work Act 1974 or any subsequent re-enactment thereof and with BS 8004 and BS 8008.

### 11.6.2. Live-saving appliances

The Contractor shall provide and maintain on the Site sufficient proper and efficient life-saving appliances. The appliances must be conspicuous and available for use at all times.

Site operatives shall be conversant with the use of safety equipment and drills shall be carried out sufficiently frequently to ensure that all necessary procedures can be correctly observed.

### 11.6.3. Diving

Diving operations shall be carried out in accordance with the Diving Operations at Work Regulations (1981), Health and Safety Executive Publication 399, and any amendments or additions thereto.

Before any diving is undertaken the Contractor shall supply the Engineer with two copies of the code of signals to be employed, and shall have a copy of the Code prominently displayed adjacent to the diving control station on the craft or structure from which any diving operation takes place.

## 11.7. Ground conditions

No responsibility is accepted by the Engineer or Employer for any opinions or conclusions given in any factual or interpretative ground investigation reports. The Contractor shall report immediately to the Engineer any circumstance which indicates that in the Contractor's opinion the ground conditions differ from those reported in or which could have been inferred from the ground investigation reports or preliminary pile results.

### 11.8.  Tolerances
#### 11.8.1.  Setting out

Marker pins for the retaining wall positions shall be set out and installed by the Contractor. Prior to installation of the retaining wall, the element positions shall be checked by the Contractor.

#### 11.8.2.  Position and verticality

The permitted deviation of element positions and alignment to the vertical is dependent upon the method of retaining wall construction and is specified in the relevant Section.

### 11.9.  Retaining wall construction method

The Contractor shall submit with his tender all relevant details of the method of constructing the retaining wall and the plant and monitoring equipment he plans to adopt. Alternative methods may be stated provided it is demonstrated that they satisfy the requirements of the Specification.

### 11.10.  Retaining wall programme

The Contractor shall submit a provisional programme for the execution of the Works at the time of tender and a detailed programme prior to commencement of the Works. He shall inform the Engineer each day of the intended programme of retaining wall construction for the following day and shall give 24 hours' notice of his intention to work outside normal hours and at weekends.

The Contractor shall ensure that during the course of the work, displacement or damage which would impair either performance or durability does not occur to completed elements.

### 11.11.  Records

The Contractor shall keep records as indicated by an asterisk in Table 11.2 for the installation of each element or as listed in Section 17 for sheet pile walls, and shall submit two signed copies of these records to the Engineer not later than noon of the next working day after the element was constructed. These records will form a record of the work.

Any unexpected driving or boring conditions shall be noted in the records.

### 11.12.  Nuisance and damage
#### 11.12.1.  Noise and disturbance

The Contractor shall carry out the work in such a manner and at such times as to minimize noise, vibration and other disturbance in order to comply with current environmental legislation.

Particular restrictions on permissible working hours are stated in the Particular Specification.

#### 11.12.2.  Damage to adjacent structures

Permissible damage criteria for adjacent structures or services are given in the Particular Specification. If in the opinion of the Contractor damage may be caused to other structures or services by his execution of the Works he shall immediately notify the Engineer. The Contractor shall submit his plans for making surveys and monitoring movements or vibration before the commencement of the Works.

The Contractor shall determine the positions of all known services and structures before commencing piling work on site.

#### 11.12.3.  Damage to completed wall elements

The Contractor shall ensure that during the course of the work, displacement or damage which would impair either performance or durability does not occur to completed wall elements.

The Contractor shall submit to the Engineer his planned sequence and timing for installing wall elements, having regard to the avoidance of damage to adjacent wall elements.

### 11.13.  Supervision and control of the Works

The Contractor shall keep upon the Works a competent site supervisor to be in charge of the retaining wall construction or installation.

The site supervisor must be experienced in the type of retaining wall construction necessitated by the Contract. A curriculum vitae for the supervisor shall be submitted prior to commencement. The whole time of the site supervisor shall be devoted to the retaining wall works. The site supervisor shall not be removed from the Works without the Engineer being notified in advance with at least one week's notice.

The Contractor shall submit to the Engineer one week prior to commencement of retaining wall works his Quality Plan for the Works. Subsequent revisions, amendments or additions shall be submitted to the Engineer prior to their implementation. Quality Assurance and Quality Control documentation shall be made available to the Engineer on request.

*Table 11.1.   Submission of information (The following submissions shall be made to the Engineer at the time stated. Detailed requirements are listed under the clause number indicated in the relevant position.)*

| Section | Item | At Tender | Prior to commencing the Works | During the Works |
|---|---|---|---|---|
| 11 | Progress Report | | | 11.3 |
| | Wall layout and design | 11.4.1 | | |
| | Retaining wall method | 11.9 | | |
| | Retaining wall programme | 11.10 | 11.10 | 11.10 |
| | Records | | | 11.11 |
| | Monitoring surveys | | 11.12.2 | |
| | Walling sequence | | 11.12.3 | |
| | Supervisor | | 11.13 | |
| | Quality Plan | | 11.3 | |
| 12 | Dimensions of panels | 12.5 | | |
| | Sequence of construction | | 12.6.1 | |
| | Stability of excavation | | | 12.6.2 |
| | Means of maintaining concrete cover | | 12.7 | |
| | Reinforcement | | 12.7 | |
| | Removal of stop ends | | | 12.10 |
| 13 | Sequence of construction | | 13.5.1 | |
| | Welding of reinforcement | | 13.6 | |
| | Means of maintaining concrete cover | | 13.6 | |
| 14 | Self-hardening slurry mix | | 14.4.2 | |
| | Sequence of construction | | 14.5.1 | |
| | Welding of reinforcement | | 14.6 | |
| | Means of maintaining concrete cover | | 14.6 | |
| 15 | Sequence of construction | | 15.5.1 | |
| | Welding of reinforcement | | 15.6 | |
| | Means of maintaining concrete cover | | 15.6 | |
| 16 | Method of construction | | 16.4.2 | |
| | Placing of King Post members | | 16.5 | |
| | Welding of reinforcement | | 16.6 | |
| | Means of maintaining concrete cover | | 16.6 | |
| 18 | Specialist testing contractor | | 18.5 | |
| | Integrity testing report | | | 18.7 |
| | Anomalous results | | | 18.8 |
| 19 | Instrumentation supplier etc. | | 19.2 | |
| | Inclinometers | | | 19.4 |
| | Calibration and data checking | | | 19.8.2 |
| | Report | | | 19.8.4 |
| | Specialist instrumentation contractor | | | 19.8.5 |
| 20 | Type of cement | | 20.2.1 | |
| | Certificates of cement conformity | | | 20.2.1 |
| | Water tests | | | 20.4.1 |
| | Concrete workability | | 20.6.2 | |
| | Evidence of ASR compliance | | 20.6.2 | |
| | Detailed information on concrete mix | | 20.6.3 | |
| | Trial mixes | | 20.7.2 | |
| | Workability of each batch | | | 20.8.2 |
| | Concrete cube tests | | | 20.8.3 |
| 21 | Support fluid mix | 21.1 | | |
| | Use and compliance | | 21.3 | |

*Specification for piling and embedded retaining walls.* Thomas Telford, London, 1996.

*Table 11.2. Records to be kept (indicated by an asterisk)*

| Data | Embedded retaining wall types | | | | |
|---|---|---|---|---|---|
| | A | B | C | D | E |
| Contract | * | * | * | * | * |
| Element reference number (location) | * | * | * | * | * |
| Element type | * | * | * | * | * |
| Nominal cross-sectional dimensions | * | * | * | * | * |
| Top and bottom of guidewall level (as appropriate) | * | * | * | - | - |
| Length of preformed element (as appropriate) | - | - | - | - | - |
| Groundwater level from direct observation | * | * | * | * | * |
| Date and time of excavation | * | * | * | * | * |
| Date of concreting | * | * | * | * | * |
| Details of material samples taken | * | * | * | * | * |
| Ground level at element position at commencement of element installation (commencing surface) | * | * | * | * | * |
| Working level on which base machine stands | * | * | * | * | * |
| Depth from ground level or guide wall, as appropriate, at element position to element toe | * | * | * | * | * |
| Toe level | * | * | * | * | * |
| Element head level as constructed | * | * | * | * | * |
| Element cut-off level | * | * | * | * | * |
| Stop end details | * | - | - | - | - |
| Length of temporary casing (as appropriate) | - | * | * | * | * |
| Length of permanent casing (as appropriate) | - | - | * | * | * |
| Soil samples taken and in-situ tests carried out during element formation or adjacent to element position | * | * | * | * | * |
| Tests on support fluid (as appropriate) | * | * | * | * | * |
| Length and details of steel reinforcement (as appropriate) | * | * | * | * | * |
| Concrete mix or grout (as appropriate) | * | * | * | * | * |
| Volume of concrete supplied to element | * | * | * | * | * |
| Graph of top of concrete or grout level vs volume placed by batch (as appropriate) | * | * | * | * | * |
| All information regarding obstructions, delays and other interruptions to the sequence of work | * | * | * | * | * |
| As constructed positional records vertical and horizontal, as required | * | * | * | * | * |
| Movements of ground and structures and services as specified | * | * | * | * | * |

**Notes**

Embedded retaining wall types:

A    Diaphragm walls
B    Hard/hard secant walls
C    Hard/soft secant walls
D    Contiguous bored pile walls
E    King Post walls

*Specification for piling and embedded retaining walls.* Thomas Telford, London, 1996.

69

# 12. Diaphragm walls

**12.1. General**

All materials and work shall be in accordance with Sections 11, 12, and 20 of this Specification, except where there may be conflict of requirements, in which case this Section shall take precedence.

**12.2. Particular Specification**

The following matters are, where appropriate, described in the Particular Specification:

(a) specified working loads
(b) performance criteria for movement under lateral loads
(c) types of cement
(d) cement replacement materials
(e) types and sizes of aggregate
(f) grades of concrete
(g) designed or prescribed mixes and maximum free water to cement ratio
(h) method of testing concrete workability
(i) grades, types and bond length of and cover to reinforcement
(j) support fluid
(k) panel dimensions (minimum thickness and maximum or minimum panel length)
(l) water stop requirements, if any
(m) water retention
(n) instrumentation
(o) temporary backfill material
(p) integrity testing
(q) disposal of excavated material
(r) other particular requirements.

**12.3. Guide walls**

The design and construction of the guide walls shall be the responsibility of the Contractor and shall take into account the actual site and ground conditions and the equipment to be used on site to ensure stability and avoid undercutting of the guide wall. Guide walls shall be constructed in reinforced concrete or other suitable materials. The minimum depth of guide wall shall be 1.0 m.

**12.4. Materials**
**12.4.1. Concrete and steel reinforcement**

Cement materials, aggregates, additives, water and steel reinforcement shall be in accordance with Section 20 of this Specification.

**12.4.2. Support fluid**

Bentonite and alternative fluid support materials, additives, mixing and testing and clean water shall be in accordance with Section 21 of this Specification.

**12.5. Dimensions of panels**

The thickness of a panel shall be not less than the specified thickness. The length of panel may be varied to suit an individual Contractor's equipment subject to any upper limits on length specified in the Particular Specification. Within these constraints the Contractor shall be responsible for selecting panel dimensions which ensure stability and that movements remain within the

*Specification for piling and embedded retaining walls.* Thomas Telford, London, 1996.

criteria set in the Particular Specification and Section 11. If in the Contractor's opinion the specified panel dimensions are not adequate to ensure stability, he shall inform the Engineer at the time of tender.

## 12.6. Excavation
### 12.6.1. Excavation near recently cast panel

Panels shall not be excavated so close to other panels which have recently been cast and which contain workable or unset concrete that a flow of concrete or instability could be induced or damage caused to any panel. The Contractor's planned sequence of construction shall be submitted prior to work commencing in accordance with Section 11 of this Specification.

### 12.6.2. Stability of the excavation

A suitable guide wall shall be used in conjunction with the method to ensure stability of the strata near ground level until concrete has been placed. During construction the level of support fluid in the excavation shall be maintained within the guide wall or stable ground so that it is not less than 1.5 m above the level of external standing groundwater at all times.

In the event of a loss of support fluid from an excavation, the Contractor shall notify the Engineer of his intended action before continuing the work.

### 12.6.3. Cleanliness of base

Prior to placing steel or concrete the Contractor shall clean the base of the excavation of as much loose, disturbed and remoulded materials as practical and in accordance with the method of construction and shall wholly or partly remove and replace support fluid while maintaining the fluid head if it does not comply with the Contractor's stated limits for support fluid prior to concreting.

## 12.7. Steel reinforcement

The number of joints in longitudinal steel bars shall be kept to a minimum.

Joints in steel reinforcement shall be such that the full strength of each bar is effective across the joint and shall be made so that there is no detrimental displacement of the reinforcement during the construction of the panel, following the guidance of BS 8110. Reinforcement shall be maintained in its correct position during concreting of the panel. Where it is made up into cages, they shall be sufficiently rigid to enable them to be handled, placed and concreted without damage. If the cage is to be welded together, welding shall be carried out to the requirements of BS 7123. Details of the procedures should be submitted prior to the commencement of the Works.

Spacers shall be designed and manufactured using durable materials which shall not lead to corrosion of the reinforcement or spalling of the concrete cover. Details of the means by which the Contractor plans to ensure the correct cover to and position of the reinforcement shall be submitted prior to commencing the Works.

The minimum projecting bond lengths required by the Particular Specification shall be observed.

The Contractor shall prepare reinforcement detail construction drawings for each panel and these shall be submitted prior to commencing the Works.

## 12.8. Placing concrete
### 12.8.1. General

The workability and method of placing of the concrete shall be such that a continuous monolithic concrete panel of the full cross-section is formed, and that the concrete in its final position is dense and homogeneous. Concrete shall be transported from the mixer to the position of the panel in such a manner that segregation of the mix does not occur.

---

Before commencement of concreting of a panel, the Contractor shall satisfy himself that the supplier will have available sufficient quantity of concrete to construct the panel in one continuous operation.

The concrete shall be placed without such interruption as would allow the previously placed batch to have achieved a stiffness which prevents proper amalgamation of the two concrete batches.

No spoil, liquid or other foreign matter shall be allowed to contaminate the concrete.

### 12.8.2. Workability of concrete

The concrete workability shall be determined using the slump or flow table in accordance with BS 1881. The slump range or target flow for concrete placed through support fluid using a tremie pipe shall be 150 mm or greater or 550 mm ±50 respectively or as specified in the Particular Specification.

### 12.8.3. Compaction

Internal vibrators shall not be used to compact concrete within a cast in place panel.

### 12.8.4. Placing concrete

The concrete shall be placed through a tremie pipe in one continuous operation. Where two or more pipes are used in the same panel simultaneously, care shall be taken to ensure that the concrete level at each pipe position is maintained nearly equal.

The hopper and pipe of the tremie shall be clean and watertight throughout. The pipe shall extend to the base of the panel and a sliding plug or barrier shall be placed in the pipe to prevent direct contact between the first charge of concrete in the tremie and the support fluid.

The pipe shall at all times penetrate the concrete which has previously been placed with a minimum embedment of 3 m and shall not be withdrawn from the concrete until completion of concreting. At all times a sufficient quantity of concrete shall be maintained within the pipe to ensure that the pressure from it exceeds that from the support fluid and workable concrete above the tremie base. The internal diameter of the pipe of the tremie shall be of sufficient size to ensure the easy flow of concrete. It shall be so designed that external projections are minimized, allowing the tremie to pass within reinforcing cages without causing damage. The internal face of the pipe of the tremie shall be free from projections.

The depth of the surface of the concrete shall be measured and the embedded length of the tremie pipe recorded at regular intervals corresponding to each batch of concrete. The depths measured and volumes placed shall be plotted immediately on a graph during the concreting process and compared with the theoretical relationship of depth against volume.

### 12.9. Tolerances
### 12.9.1. Guide wall

The finished internal face of the guide wall closest to any subsequent main excavation shall be vertical to within a tolerance of 1 in 200 and the top edge of the wall shall represent the reference line. There shall be no ridges or abrupt changes on the face and its variation from its specified position shall not exceed ±15 mm in 3 m.

The minimum clear distance between the guide walls shall be the specified diaphragm wall thickness plus 25 mm and the maximum distance shall be the specified diaphragm wall thickness plus 50 mm.

### 12.9.2. Diaphragm wall

At cut-off level the maximum deviation of the centre-line of each panel from the specified position shall be 15 mm and an additional tolerance of 8 mm for each 1.0 m that the cut-off level is below the top of the guide wall shall be permitted unless otherwise stated in the Particular Specification.

The exposed wall face and the ends of panels shall be vertical within a tolerance of 1:120 or as specified in the particular specification. An additional tolerance of 100 mm will be allowed for concrete protrusions resulting from cavities formed by overbreak in the ground. Where very soft clay layers or peat layers are anticipated or obstructions are to be removed during trench excavation, an additional overbreak tolerance shall be stated in the Particular Specification.

### 12.9.3. Recesses

Where recesses are to be formed by inserts in the wall, the vertical tolerance shall be that of Clause 12.9.4 and the horizontal tolerance shall be that of Clause 12.9.4 plus the horizontal tolerance resulting from Clause 12.9.2.

### 12.9.4. Steel reinforcement

The longitudinal tolerance of the cage head at the top of the guide wall measured along the excavation shall be ±75 mm.

The vertical tolerance of the cage head measured relative to the guide wall shall be +150/−50 mm. The reinforcement shall be maintained in position during concreting of a panel.

### 12.10. Temporary stop-ends in diaphragm panels

Temporary stop-ends shall be of the length, thickness and quality of material adequate for the purpose of preventing water and soil from entering the panel excavations.

Each temporary stop-end shall be straight and true throughout. The external surface shall be clean and smooth, i.e. free from distortions that may affect panel integrity during removal of the temporary stop-end.

Stop-ends shall be rigid and adequately restrained to prevent horizontal movement during concreting. The Contractor shall notify the Engineer prior to the removal of each stop-end.

### 12.11. Concrete level

If the cut-off level for the panel is less than 1 m below the top level of the guide walls, uncontaminated concrete shall be brought to the top of the guide walls. If the cut-off level is greater than 1 m below the top level of the guide walls, concrete shall be brought to 1 m above the cut-off level specified, with a tolerance of ±150 mm. An additional tolerance of +150 mm over the above tolerances shall be permitted for each 1.0 m of depth by which the cut-off level is below the top of the guide wall.

Where more than one tremie pipe is used the concrete shall be brought up to 1 m above the cut-off level specified with a tolerance of ±250 mm.

### 12.12. Temporary backfilling above panel casting level

After each panel has been cast, any empty excavation remaining shall be protected and shall be carefully backfilled as soon as possible, with material in accordance with the Particular Specification. Prior to backfilling, panels shall be clearly marked and fenced off so as not to cause a safety hazard.

### 12.13. Water retention

The Contractor shall be responsible for the repair of any joint, defect or panel where on exposure of the wall visible running water leaks are found which would result in leakage per individual square metre in excess of that stated in the Particular Specification. Any

*Specification for piling and embedded retaining walls.* Thomas Telford, London, 1996.

73

leak which results in a flow emanating from the surface of the retaining wall shall be sealed.

## 12.14. Instrumentation

Movement of the wall, the surrounding ground and existing structures shall be monitored in accordance with the Particular Specification and Section 19 of this Specification.

# 13. Hard/hard secant pile walls

**13.1. General**

All materials and work shall be in accordance with Sections 3, 4, 11, 13 and 20 of this Specification, except where there may be conflict of requirements, in which case this Section shall take precedence.

**13.2. Particular Specification**

The following matters are, where appropriate, described in the Particular Specification:

- (*a*) specified working loads (if any)
- (*b*) performance criteria for movement under lateral loads
- (*c*) types of cement
- (*d*) cement replacement materials
- (*e*) types and sizes of aggregate
- (*f*) grades of concrete
- (*g*) designed or prescribed mixes and maximum free water to cement ratio
- (*h*) method of testing concrete workability
- (*i*) grades, types and bond length of, and cover to reinforcement
- (*j*) support fluid
- (*k*) panel dimensions (minimum thickness and maximum or minimum panel length)
- (*l*) water stop requirements, if any
- (*m*) water retention
- (*n*) instrumentation
- (*o*) temporary backfill material
- (*p*) integrity testing
- (*q*) disposal of excavated material
- (*r*) other particular requirements.

**13.3. Guide walls**

The design and construction of guide walls, if the use of guide walls is specified in the Particular Specification or is required by the Contractor, shall be the responsibility of the Contractor and shall take into account the actual site and ground conditions and the equipment to be used on site to ensure stability and avoid undercutting as appropriate.

Guide walls shall be constructed in reinforced concrete or other suitable materials. The minimum depth of guide wall shall be 0.5 m, and the minimum shoulder width shall be 0.3 m for walls in reinforced concrete.

**13.4. Materials**
*13.4.1. Concrete and steel reinforcement*

Cement materials, aggregates, additives, water and steel reinforcement shall be in accordance with Section 20 of this Specification.

*13.4.2. Support fluid*

Where support fluid is used to maintain the stability of the pile bore it shall be in accordance with Section 21.

*Specification for piling and embedded retaining walls*. Thomas Telford, London, 1996.

75

## 13.5. Boring
### 13.5.1. Boring near recently cast piles

Piles shall not be bored so close to other piles which have recently been cast and which contain workable or partially set concrete that a flow of concrete or instability could be induced or damage caused to any installed piles. The Contractor's sequence of construction shall be submitted prior to work commencing in accordance with Section 11 of this Specification.

### 13.5.2. Stability of bore

Temporary casing shall be used in unstable ground with or without support fluid to maintain the stability of the bore unless continuous flight augers are used. The process of advancing the bore and the temporary casing shall be such that soil is not drawn into the bore from outside the area of the pile and cavities are not created outside the temporary casing.

### 13.5.3. Cleanliness of pile base

Prior to placing steel or concrete, the Contractor shall remove from the base of the pile as much remoulded or loose disturbed materials as practical and in accordance with the method of construction.

### 13.5.4. Cleaning of support fluid

Prior to placing steel or concrete, any support fluid including water shall be wholly or partly removed and replaced while maintaining the fluid head if it does not comply with the Contractor's stated limits for support fluid prior to concreting.

## 13.6. Steel reinforcement

The number of joints in longitudinal steel bars shall be kept to a minimum. Joints in steel reinforcement shall be such that the full strength of each bar is effective across the joint and shall be made so that there is no detrimental displacement during the construction of the pile, following the guidance of BS 8110.

Reinforcement shall be maintained in its correct position during concreting of the wall element. Where it is made up into cages, they shall be sufficiently rigid to enable them to be handled, placed and concreted without damage. If the cage is to be welded together, welding shall be carried out to the requirements of BS 7123. Details of the procedures shall be submitted prior to the commencement of the Works.

Spacers shall be designed and manufactured using durable materials which shall not lead to corrosion of the reinforcement or spalling of the concrete cover. Details of the means by which the Contractor plans to ensure the correct cover to and position of the reinforcement shall be submitted prior to commencement of the Works.

The minimum projecting bond lengths required by the Particular Specification shall be maintained.

For continuous flight auger piles steel reinforcement cages shall be inserted into the pile shaft immediately after completion of concreting. The method of insertion of reinforcement cages shall avoid distortion or damage to the steel reinforcement and ensure accurate positioning of the cage within the pile shaft.

## 13.7. Placing and compacting concrete
### 13.7.1. General

The workability and method of placing of the concrete shall be such that a continuous monolithic concrete shaft of the full cross-section is formed, and that the concrete in its final position is dense and homogeneous. Concrete shall be transported from the mixer to the position of the pile in such a manner that it does not cause segregation of the mix.

Before commencement of concreting of a pile, the Contractor shall satisfy himself that the supplier will have available a sufficient quantity of concrete to construct the pile in one continuous operation.

Each batch of concrete in a pile shall be placed before the previous batch has achieved a stiffness which prevents proper amalgamation of the two concrete batches. Removal of temporary casings, when used, shall be completed before the concrete within the casing loses its workability.

For continuous flight auger piles injection of concrete shall continue up to the commencing surface so that free discharge from the delivery pipe can be observed to be taking place, and reinforcement placed after concreting.

No spoil, liquid or other foreign matter shall be allowed to contaminate the concrete.

### 13.7.2. Workability of concrete

The concrete workability shall be determined using the slump or flow table in accordance with BS 1881 and as specified in the Particular Specification. The slump or flow shall be measured at the time of discharge into the pile boring and shall be in accordance with the limits shown in Table 3.1.

### 13.7.3. Compaction

Internal vibrators shall not be used to compact concrete.

### 13.7.4. Placing concrete in dry boreholes

The concrete shall be placed through a hopper attached to a length of tremie pipe and in such a way that the flow is directed and does not hit reinforcing bars or the side of the bore.

The tremie pipe shall be at least 3 m long.

This Clause is not applicable to piles constructed using continuous flight augers.

### 13.7.5. Placing concrete under water or support fluid

Concrete to be placed under water or support fluid shall be placed through a tremie pipe in one continuous operation.

The hopper and pipe of the tremie shall be clean and watertight throughout. The pipe shall extend to the base of the boring and a sliding plug or barrier shall be placed in the pipe to prevent direct contact between the first charge of concrete in the pipe of the tremie and the water or support fluid. The pipe shall at all times penetrate the concrete which has previously been placed with a minimum penetration of 3 m and shall not be withdrawn from the concrete until completion of concreting.

At all times a sufficient quantity of concrete shall be maintained within the tremie pipe to ensure that the pressure from it exceeds that from the water or support fluid and workable concrete above the tremie base.

The internal diameter of the pipe of the tremie shall be of sufficient size to ensure the easy flow of concrete. It shall be so designed that external projections are minimized, allowing the tremie to pass through reinforcing cages without causing damage. The internal face of the pipe of the tremie shall be free from projections.

The depths to the surface of the concrete shall be measured and the embedded length of the tremie pipe recorded at regular intervals corresponding to the placing of each load of concrete. The depth measured and volumes of concrete placed shall be plotted immediately on a graph during the concreting process and compared with the theoretical relationship of depth against volume.

### 13.7.6. Placing concrete in continuous flight auger piles

The placing of concrete in continuous flight auger piles shall be in accordance with Section 4.

### 13.7.7. Time period for excavation and placing concrete

The time period after a pile is excavated and before the concrete is placed shall not exceed 12 hours unless otherwise specified in the Particular Specification.

If temporary casing which has the same diameter as the pile bore is used, this period shall start when excavation below the temporary casing commences.

### 13.8. Tolerances
### 13.8.1. Guide wall

The finished internal face of the guide wall closest to any subsequent main excavation shall be vertical to a tolerance of 1 in 200 and shall represent the reference line. There shall be no ridges on the face and the centre line of the guide wall shall not deviate from its specified position by more than ±15 mm in 3 m.

The minimum clear distance between the guide walls shall be the pile diameter plus 25 mm and the maximum distance shall be the pile diameter plus 50 mm.

### 13.8.2. Secant piles

At cut-off level the maximum permitted deviation of the pile centre from the centre point shown on the setting out drawings shall be 25 mm in any direction except that an additional tolerance of 5 mm shall be permitted for each additional 1.0 m that the cut-off level is below the top of the guide wall unless otherwise specified in the Particular Specification.

In the case of construction by continuous flight augers the additional tolerance shall be 8 mm for each 1.0 m below top of guide wall.

The exposed face of the pile shall be vertical within a tolerance of 1:200 unless otherwise stated in the Particular Specification.

An additional tolerance of 100 mm will be allowed for concrete protrusions resulting from cavities formed by overbreak in the ground. Where very soft clay layers or peat layers are anticipated or obstructions are to be removed prior to or during boring, an additional overbreak tolerance shall be stated in the Particular Specification.

### 13.8.3. Recesses

If recesses in the form of box-outs are to be formed within a pile shaft, the vertical tolerance shall be in accordance with Clause 13.8.4 and rotational tolerance shall be 10 degrees.

### 13.8.4. Steel reinforcement

A vertical tolerance of +150 mm/−50 mm shall generally be permitted on the level of steel projecting from a pile cut-off.

### 13.9. Concrete level

No pile shall be cast with its final cut-off level below standing water level unless the Contractor's method of construction includes measures to prevent inflow of water that may locally reduce the pile's cross-sectional area (necking) or contaminate the concrete as temporary casing is withdrawn. The standing water level will be treated as the cut-off level for the purpose of calculating the casting level tolerance.

For piles cast in dry bores using temporary casing, and for piles cast under water the casting level above cut-off level shall be within the casting tolerances shown in Table 3.2 except for continuous flight auger piles which shall be cast to ground level.

### 13.10. Backfilling of empty bore

Where concrete is not brought to the top of the guide wall, the empty pile bore shall be backfilled as soon as possible with material in accordance with the Particular Specification. Prior to backfilling the bore shall be clearly marked and fenced off so as not to cause a safety hazard.

### 13.11. Water retention

The Contractor shall be responsible for the repair of any joint, defect or pile where on exposure of the wall visible running water leaks are found which would result in leakage per individual square metre in excess of that stated in the Particular Specification. Any leak which results in water flow emanating from the surface of the retaining wall shall be sealed.

### 13.12. Instrumentation

Movement of the wall, the surrounding ground and existing structures shall be monitored in accordance with the Particular Specification and Section 19 of this Specification.

*Specification for piling and embedded retaining walls*. Thomas Telford, London, 1996.

79

# 14. Hard/soft secant pile walls

**14.1. General**

All materials and work shall be in accordance with Sections 3, 4, 11, 14 and 20 of this Specification, except where there may be conflict of requirements, in which case this Section shall take precedence.

**14.2. Particular Specification**

The following matters are, where appropriate, described in the Particular Specification:

(a) specified working loads (if any)
(b) performance criteria for movement under lateral loads
(c) types of cement
(d) cement replacement materials
(e) types and sizes of aggregate
(f) grades of concrete or alternative pile mixes
(g) designed or prescribed mixes and maximum free water to cement ratio
(h) method of testing concrete workability
(i) grades, types and bond length of, and cover to reinforcement
(j) support fluid
(k) requirements for self-hardening slurry mix
(l) pile diameters
(m) pile spacings and overlap at commencing level
(n) water retention
(o) instrumentation
(p) temporary backfill material
(q) integrity testing
(r) disposal of excavated material
(s) other particular requirements.

**14.3. Guide walls**

The design and construction of guide walls, if the use of guide walls is specified in the Particular Specification or is required by the Contractor, shall be the responsibility of the Contractor and shall take into account the actual site and ground conditions and the equipment to be used on site to ensure stability and avoid undercutting as appropriate.

Guide walls shall be constructed in reinforced concrete or other suitable materials. The minimum depth of guide wall shall be 0.5 m and the minimum shoulder width shall be 0.3 m for walls in reinforced concrete.

**14.4. Materials**
**14.4.1. Concrete and steel reinforcement**

Cement materials, aggregates, additives, water and steel reinforcement shall be in accordance with Section 20 of this Specification.

**14.4.2. Self-hardening slurry mixes**

The Contractor shall submit details for the self-hardening slurry mix (soft pile) proportions to be used prior to commencing the Works.

Cement materials, aggregates (if used), additives and water shall be in accordance with Section 20 of this Specification.

*Specification for piling and embedded retaining walls*. Thomas Telford, London, 1996.

Trial mixes shall be prepared for each mix unless there are existing data showing that the mix proportions and method of manufacture will produce hardened material of the strength, permeability, shrinkage properties and durability required, having adequate workability for compaction by the method to be used in placing. The performance requirements are set out in the Particular Specification. Limits on workability and time for placement are included. The requirements for testing will be set out in the Particular Specification and samples shall be tested in a NAMAS accredited laboratory for those tests.

Where a trial mix is required after commencement of the Works the above procedure shall be adopted. The workability of each batch of a trial mix shall be determined by the method as specified in the Particular Specification.

No variations outside the limits set out in the proportions shall be made nor in the original source of the materials without demonstrating compliance with this Specification.

Self-hardening mixes shall be checked for compliance with the mix proportions. Cylindrical samples shall be made at the rate of four samples for each 50 m$^3$ of self hardening slurry or part thereof in each day's work. Testing will be in accordance with the requirements of the Particular Specification and carried out in a NAMAS accredited laboratory for those tests. The results shall comply with the Particular Specification.

The Contractor shall keep a detailed record of the results of all tests on self-hardening mixes and their ingredients. Each test shall be clearly identified with the pile to which it relates and the date it was carried out.

### 14.4.3. Support fluid

Where support fluid is used to maintain the stability of the pile bore it shall be in accordance with Section 21.

## 14.5. Boring
### 14.5.1. Boring near recently cast piles

Piles shall not be bored so close to recently cast piles which contain workable or partially set concrete piles or self hardening mixes that a flow of concrete or instability could be induced or damage caused to any installed piles. The Contractor's sequence of construction shall be submitted prior to work commencing in accordance with Section 11 of this Specification.

### 14.5.2. Stability of bore

Temporary casing shall be used in unstable ground with or without support fluid to maintain the stability of the bore unless continuous flight augers are used. The process of advancing the bore and the temporary casing shall be such that soil is not drawn into the bore from outside the area of the pile and cavities are not created outside the temporary casing.

### 14.5.3. Cleanliness of pile base

Prior to placing steel or concrete, the Contractor shall remove from the base of the pile as much loose disturbed and remoulded materials as practical and in accordance with the method of construction.

### 14.5.4. Cleaning of support fluid

Prior to placing steel or concrete, any support fluid including water shall be wholly or partly removed and replaced while maintaining the fluid head if it does not comply with the Contractor's limits for support fluid prior to concreting.

## 14.6. Steel reinforcement

The number of joints in longitudinal steel bars shall be kept to a minimum. Joints in reinforcement shall be such that the full strength of the bar is effective across the joint and shall be made so that there is no detrimental displacement during the construction of the pile, following the guidance of BS 8110.

Reinforcement shall be maintained in its correct position during concreting of the pile. Where it is made up into cages, they shall be sufficiently rigid to enable them to be handled, placed and concreted without damage. If the cage is to be welded together, welding shall be carried out to the requirements of BS 7123. Details of the procedures shall be submitted prior to the commencement of the Works.

Spacers shall be designed and manufactured using durable materials which shall not lead to corrosion of the reinforcement or spalling of the concrete cover. Details of the means by which the Contractor plans to ensure the correct cover to and position of the reinforcement shall be submitted prior to commencement of the Works.

The minimum projecting bond lengths required by the Particular Specification shall be maintained.

For continuous flight auger piles steel reinforcement cages shall be inserted into the pile shaft immediately after completion of concreting. The method of insertion of reinforcement cages shall avoid distortion or damage to the steel reinforcement and ensure accurate positioning of the cage within the pile shaft.

## 14.7. Placing and compacting concrete
### 14.7.1. General

The workability and method of placing of the concrete shall be that a continuous monolithic concrete shaft of the full cross-section is formed, and that the concrete in its final position is dense and homogeneous. Concrete shall be transported from the mixer to the position of the pile in such a manner that it does not cause segregation of the mix.

Before commencement of concreting of a pile, the Contractor shall satisfy himself that the supplier shall have available a sufficient quantity of concrete to construct the pile in one continuous operation.

Each batch of concrete in a pile shall be placed before the previous batch has achieved a stiffness which prevents proper amalgamation of the two concrete batches. Removal of temporary casings, when used, shall be completed before the concrete within the casing loses its workability.

For continuous flight auger piles injection of concrete shall continue up to the commencing surface so that free discharge from the delivery pipe can be observed to be taking place, and reinforcement placed after concreting.

No spoil, liquid or other foreign matter shall be allowed to contaminate the concrete.

### 14.7.2. Workability of concrete

The concrete workability shall be determined using the slump or flow table in accordance with BS 1881 and as specified in the Particular Specification. The slump or flow shall be measured at the time of discharge into the pile boring and shall be in accordance with the limits shown in Table 3.1.

### 14.7.3. Compaction

Internal vibrators shall not be used to compact concrete.

### 14.7.4. Placing concrete in dry boreholes

The concrete shall be placed through a hopper attached to a length of tremie pipe and in such a way that the flow is directed and does not hit reinforcing bars or the side of the bore. The tremie pipe shall be at least 3 m long.

This Clause is not applicable to piles constructed using continuous flight augers.

### 14.7.5. Placing concrete under water or support fluid

Concrete to be placed under water or support fluid shall be placed through a tremie pipe in one continuous operation.

The hopper and pipe of the tremie shall be clean and watertight throughout. The pipe shall extend to the base of the boring and a sliding plug or barrier shall be placed in the pipe to prevent direct contact between the first charge of concrete in the pipe of the tremie and the water or support fluid. The pipe shall at all times penetrate the concrete which has previously been placed with a minimum penetration of 3 m and shall not be withdrawn from the concrete until completion of concreting.

At all times a sufficient quantity of concrete shall be maintained within the tremie pipe to ensure that the pressure from it exceeds that from the water or support fluid and workable concrete above the tremie base.

The internal diameter of the pipe of the tremie shall be of sufficient size to ensure the easy flow of concrete. It shall be so designed that external projections are minimized, allowing the tremie to pass through reinforcing cages without causing damage. The internal face of the pipe of the tremie shall be free from projections.

The depths to the surface of the concrete shall be measured and the embedded length of the tremie pipe recorded at regular intervals corresponding to the placing of each batch of concrete. The depth measured and volumes of concrete placed shall be plotted immediately on a graph during the concreting process and compared with the theoretical relationship of depth against volume.

### 14.7.6. Placing concrete in continuous flight auger piles

The placing of concrete in continuous flight auger piles shall be in accordance with Section 4.

### 14.7.7. Time period for excavation and placing concrete

The time period after a pile is excavated and before the concrete is placed shall not exceed 12 hours unless otherwise specified in the Particular Specification.

If temporary casing which has the same diameter as the pile bore is used, this period shall start when excavation below the temporary casing commences.

### 14.8. Placing and compacting self-hardening slurry mixes
### 14.8.1. General

The method of placing and the workability of a self-hardening slurry mix shall be such that a continuous monolithic shaft of the full cross-section is formed and that it shall be homogeneous in its final position.

Before commencing the filling of the pile the Contractor shall plan and reasonably demonstrate that a sufficient quantity of self-hardening mix is available to construct the pile in one continuous operation.

Removal of temporary casing, when used, shall be completed before the self-hardening mix within the casing loses its workability.

For continuous flight auger piles injection of self-hardening mix shall continue up to the commencing surface so that free discharge from the delivery pipe can be observed to be taking place.

No spoil, liquid or other foreign matter shall be allowed to contaminate the mix.

### 14.8.2. Workability of self-hardening mix

Self-hardening mixes shall be coherent and of a workability such that when in its final position and after all constructional procedures in forming the pile have been completed it shall remain sufficiently workable.

### 14.8.3. Placing self-hardening mixes

The self-hardening mix shall be placed using methods appropriate to the composition of the mix. These may include placing through a hopper attached to a length of tremie pipe and tremie methods in which case placement shall be in accordance with Clauses 14.7.6 and 14.7.7. Other self-hardening mixes use proprietary methods which shall be set out by the Contractor in his method of construction. Continuous flight auger piles shall use an appropriate pump which shall be kept charged continuously throughout pile construction.

### 14.8.4. Time period for excavation and placing self-hardening mix

The time period after a pile is excavated and before the self-hardening mix is placed shall not exceed 12 hours unless otherwise permitted in the Particular Specification. If temporary casing is used, this period shall start when excavation below the temporary casing commences.

### 14.9. Tolerances
### 14.9.1. Guide wall

The finished face of the guide wall closest to any subsequent main excavation shall be vertical to a tolerance of 1 in 200 and shall represent the reference line. There shall be no ridges on the face and the centre-line of the guide wall shall not deviate from its specified position by more than ±15 mm in 3 m.

The minimum clear distance between the guide walls shall be the pile diameter plus 25 mm and the maximum distance shall be the pile diameter plus 50 mm.

### 14.9.2. Secant piles

At cut-off level the maximum permitted deviation of the pile centre from the centre point shown on the setting out drawings shall be 25 mm in any direction except that an additional tolerance of 10 mm shall be permitted for each additional 1.0 m that the cut-off level is below the top of the guide wall unless otherwise specified in the Particular Specification.

In the case of construction by continuous flight augers, the additional tolerance shall be 8 mm for each 1.0 m below top of guide wall.

The exposed face of the pile shall be vertical within a tolerance of 1:100 in any direction unless otherwise stated in the Particular Specification.

An additional tolerance of 100 mm will be allowed for concrete protrusions resulting from cavities formed by overbreak in the ground. Where very soft clay layers or peat layers are anticipated or obstructions are to be removed prior to or during boring, an additional overbreak tolerance will be stated in the Particular Specification.

### 14.9.3. Recesses

If recesses in the form of box-outs are to be formed within a pile shaft, the vertical tolerance shall be in accordance with Clause 14.9.4 and the rotational tolerance shall be 10 degrees.

### 14.9.4. Steel reinforcement

A vertical tolerance of +150/−50 mm shall generally be permitted on the level of steel projecting from a pile cut-off.

### 14.10. Concrete level

No pile shall be cast with its final cut-off level below standing water level unless the Contractor's method of construction includes measures to prevent inflow of water that may locally reduce the pile's cross-sectional area (necking) or contaminate the concrete as temporary casing is withdrawn. The standing water level will be treated as the cut-off level for the purpose of calculating the casting level tolerance.

For piles cast in dry bores using temporary casing, and for piles cast under water or support fluid the casting level above cut-off level shall be within the tolerance above the cut-off level shown in Table 3.2 except for continuous flight auger piles which shall be cast to ground level.

## 14.11. Backfilling of empty bore

Where concrete is not brought to the top of the guide wall, the empty pile bore shall be backfilled as soon as possible with material in accordance with the Particular Specification. Prior to backfilling, bores shall be clearly marked and fenced off so as not to cause a safety hazard.

## 14.12. Water retention

The Contractor shall be responsible for the repair of any joint, defect or pile where on exposure of the wall visible running water leaks are found which would result in leakage per individual square metre in excess of that stated in the Particular Specification. Any leak which results in water flow emanating from the surface of the retaining wall shall be sealed.

## 14.13. Instrumentation

Movement of the wall, the surrounding ground and existing structures shall be monitored in accordance with the Particular Specification and Section 19 of this Specification.

# 15.    Contiguous bored pile walls

### 15.1.  General

All materials and work shall be in accordance with Sections 3, 4, 11, 15 and 20 of this Specification, except where there may be conflict of requirements in which case this Section shall take precedence.

### 15.2.  Particular Specification

The following matters are, where appropriate, described in the Particular Specification:

(a)   specified working loads (if any)
(b)   performance criteria for movement under lateral loads
(c)   types of cement
(d)   cement replacement materials
(e)   types and sizes of aggregate
(f)   grades of concrete
(g)   designed or prescribed mixes and maximum free water to cement ratio
(h)   method of testing concrete workability
(i)   grades, types and bond length of, and cover to reinforcement
(j)   support fluid
(k)   pile diameters
(l)   pile spacing
(m)   additional measures for water retention
(n)   instrumentation
(o)   temporary backfill material
(p)   integrity testing
(q)   disposal of excavated material
(r)   other particular requirements.

### 15.3.  Guide walls

The design and construction of guide walls, if the use of guide walls is specified in the Particular Specification or is required by the Contractor, shall be the responsibility of the Contractor and shall take into account the actual site and ground conditions and the equipment to be used on site to ensure stability and avoid undercutting as appropriate.

Guide walls shall be constructed in reinforced concrete or other suitable materials. The minimum depth of guide wall shall be 0.5 m and the minimum shoulder width shall be 0.3 m for walls in reinforced concrete.

### 15.4.  Materials
#### 15.4.1.  Concrete and steel reinforcement

Cement materials, aggregates, additives, water and steel reinforcement shall be in accordance with Section 20 of this Specification.

#### 15.4.2.  Support fluid

Where support fluid is used to maintain the stability of the pile bore it shall be in accordance with Section 21 of this Specification.

### 15.5.  Boring
#### 15.5.1.  Boring near recently cast piles

Piles shall not be bored so close to recently cast piles which contain workable or partially set concrete that a flow of concrete or instability could be induced or damage caused to any installed piles.

*Specification for piling and embedded retaining walls.* Thomas Telford, London, 1996.

The Contractor's planned sequence of construction shall be submitted prior to work commencing in accordance with Section 11 of this Specification.

### 15.5.2. Stability of bore

Temporary casings shall be used in unstable ground with or without support fluid to maintain the stability of the bore unless continuous flight augers are used. The process of advancing the bore and the temporary casing shall be such that soil is not drawn into the bore from outside the area of the pile and cavities are not created outside the temporary casing.

Temporary casings shall be extracted immediately on completion of concreting.

### 15.5.3. Cleanliness of pile base

Prior to placing steel or concrete, the Contractor shall remove from the base of the pile as much remoulded or loose disturbed materials as practical and in accordance with the method of construction.

### 15.5.4. Cleaning of support fluid

Prior to placing steel or concrete, any support fluid including water shall be wholly or partly removed and replaced while maintaining the fluid head if it does not comply with the Contractor's limits for support fluid prior to concreting.

## 15.6. Steel reinforcement

The number of joints in longitudinal steel bars shall be kept to a minimum. Joints in reinforcement shall be such that the full strength of the bar is effective across the joint and shall be made so that there is no relative displacement during the construction of the pile, following the guidance of BS 8110.

Reinforcement shall be maintained in its correct position during concreting of the wall element. Where it is made up into cages, they shall be sufficiently rigid to enable them to be handled, placed and concreted without damage. If the cage is to be welded together, welding shall be carried out to the requirements of BS 7123. Details of the procedures shall be submitted prior to commencement of the Works.

Spacers shall be designed and manufactured using durable materials which shall not lead to corrosion of the reinforcement or spalling of the concrete cover. Details of the means by which the Contractor plans to ensure the correct cover to and position of the reinforcement shall be submitted prior to commencing the Works.

The minimum projecting bond lengths required by the Particular Specification shall be maintained.

For continuous flight auger piles steel reinforcement cages shall be inserted into the pile shaft immediately after completion of concreting. The method of insertion of reinforcement cages shall avoid distortion or damage to the steel reinforcement and ensure accurate positioning of the cage within the pile shaft.

## 15.7. Placing and compacting concrete
### 15.7.1. General

The workability and method of placing of the concrete shall be such that a continuous monolithic concrete shaft of the full cross-section is formed, and that the concrete in its final position is dense and homogeneous. Concrete shall be transported from the mixer to the position of the pile in such a manner that it does not cause segregation of the mix.

Before commencement of concreting of the pile the Contractor shall satisfy himself that the supplier will have available a sufficient quantity of concrete to construct the pile in one continuous operation.

Each batch of concrete in a pile shall be placed before the previous batch has achieved a stiffness which prevents proper

*Specification for piling and embedded retaining walls.* Thomas Telford, London, 1996.

87

amalgamation of the two concrete batches. Removal of temporary casings, where used, shall be completed before the concrete within the casing loses its workability.

For continuous flight auger piles injection of concrete shall continue up to the commencing surface so that free discharge from the delivery pipe can be observed to be taking place, and reinforcement placed after concreting.

No spoil, liquid or other foreign matter shall be allowed to contaminate the concrete.

### 15.7.2. Workability of concrete

The concrete workability shall be determined using the slump or flow table in accordance with BS 1881 and as specified in the Particular Specification. The slump or flow shall be measured at the time of discharge into the pile boring and shall be in accordance with the limits shown in Table 3.1.

### 15.7.3. Compaction

Internal vibrators shall not be used to compact concrete.

### 15.7.4. Placing concrete in dry bores

The concrete shall be poured through a hopper attached to a length of tremie pipe and in such a way that the flow is directed and does not hit reinforcing bars or the side of the bore. The tremie pipe shall be at least 3 m long.

This Clause is not applicable to piles constructed using continuous flight augers.

### 15.7.5. Placing concrete under water or support fluid

Concrete to be placed under water or support fluid shall be placed through a tremie pipe in one continuous operation.

The hopper and pipe of the tremie shall be clean and watertight throughout. The pipe shall extend to the base of the boring and a sliding plug or barrier shall be placed in the pipe to prevent direct contact between the first charge of concrete in the pipe of the tremie and the water or support fluid. The pipe shall at all times penetrate the concrete which has previously been placed with a minimum penetration of 3 m and shall not be withdrawn from the concrete until completion of concreting.

At all times a sufficient quantity of concrete shall be maintained within the tremie pipe to ensure that the pressure from it exceeds that from the water or support fluid and workable concrete above the tremie base.

The internal diameter of the pipe of the tremie shall be of sufficient size to ensure the easy flow of concrete. It shall be so designed that external projections are minimized, allowing the tremie to pass through reinforcing cages without causing damage. The internal face of the pipe of the tremie shall be free from projections.

The depths to the surface of the concrete shall be measured and the embedded length of the tremie pipe recorded at regular intervals corresponding to the placing of each batch of concrete. The depth measured and volumes of concrete placed shall be plotted immediately on a graph during the concreting process and compared with the theoretical relationship of depth against volume.

### 15.7.6. Placing concrete in continuous flight auger piles

The placing of concrete in continuous flight auger piles shall be in accordance with Section 4.

### 15.7.7. Time period for excavation and placing concrete

The time period after a pile is excavated and before the concrete is placed shall not exceed 12 hours unless otherwise specified in the Particular Specification.

If temporary casing which has the same diameter as the pile bore is used, this period shall start when excavation below the temporary casing commences.

### 15.8. Tolerances
#### 15.8.1. Guide wall

The finished internal face of the guide wall closest to any subsequent main excavation shall be vertical to a tolerance of 1 in 200 and shall represent the reference line. There shall be no ridges on the face and the centre-line of the guide wall shall not deviate from its specified position by more than ±15 mm in 3 m.

The minimum clear distance between the guide walls shall be the pile diameter plus 25 mm and the maximum distance shall be the pile diameter plus 50 mm.

#### 15.8.2. Contiguous piles

At cut-off level the maximum permitted deviation of the pile centre from the centre point shown on the setting out drawings shall be 50 mm in any direction except that an additional tolerance of 10 mm shall be permitted for each additional 1.0 m that the cut-off level is below working level or top of the guide wall unless otherwise specified in the Particular Specification.

In the case of construction by continuous flight augers the additional tolerance shall be 8 mm for each 1.0 m below the top of the guide wall.

The exposed face of the pile shall be vertical within a tolerance of 1:75 in any direction unless otherwise stated in the Particular Specification.

An additional tolerance of 100 mm will be allowed for concrete protrusions resulting from cavities formed by overbreak in the ground. Where very soft clay layers or peat layers are anticipated or obstructions are to be removed prior to or during boring, an additional overbreak tolerance shall be stated in the Particular Specification.

#### 15.8.3. Recesses

If recesses in the form of box-outs are to be formed within a pile shaft, vertical tolerance shall be in accordance with Clause 15.8.4 and rotational tolerance shall be 10 degrees.

#### 15.8.4. Steel reinforcement

A vertical tolerance of +150/−50 mm shall generally be permitted on the level of steel projecting from a pile cut-off.

### 15.9. Concrete level

No pile shall be cast with its final concreted level below standing water level unless the Contractor's method of construction includes measures to prevent inflow of water that may locally reduce the pile's cross-sectional area (necking) or contaminate the concrete as the temporary casing is withdrawn. The standing water level will be treated as the cut-off level for the purpose of calculating the casting level tolerance.

For piles cast in dry bores and for piles cast under water or support fluid the casting level above cut-off level shall be within the tolerance above the cut-off level shown in Table 3.2 except for continuous flight auger piles which shall be cast to ground surface.

### 15.10. Backfilling of empty bore

Where the concrete is below the working level, the empty pile bore shall be backfilled as soon as possible with material in accordance with the Particular Specification. Prior to backfilling, bores shall be clearly marked and fenced off so as not to cause a safety hazard.

### 15.11. Instrumentation

Movement of the wall, the surrounding ground and existing structures shall be monitored in accordance with the Particular Specification and Section 19 of this Specification.

# 16.  King Post walls

## 16.1.  General

All materials and work shall be in accordance with Sections 3, 4, 11, 16 and 20 of this Specification, except where there may be conflict of requirements, in which case this Section shall take precedence.

## 16.2.  Particular Specification

The following matters are, where appropriate, described in the Particular Specification:

(*a*)  specified working loads (if any)
(*b*)  performance criteria for movement under lateral loads
(*c*)  requirements of King Post members
(*d*)  types of cement
(*e*)  cement replacement materials
(*f*)  types and sizes of aggregate
(*g*)  grades of concrete
(*h*)  method of testing concrete workability
(*i*)  designed or prescribed mixes and maximum free water to cement ratio
(*j*)  grades, types and bond length of, and cover to reinforcement
(*k*)  support fluid
(*l*)  pile diameters
(*m*)  pile spacing
(*n*)  instrumentation
(*o*)  temporary backfill material
(*p*)  disposal of excavated material
(*q*)  requirements and specification for logging
(*r*)  whether columns can be plunged
(*s*)  other particular requirements.

## 16.3.  Materials
### 16.3.1.  Concrete and steel reinforcement

Cement materials, aggregates, additives, water and steel reinforcement shall be in accordance with Section 20 of this Specification.

### 16.3.2.  Support fluid

Where support fluid is used to maintain the stability of the pile bore it shall be in accordance with Section 21.

### 16.3.3.  Structural steelwork

The Contractor shall provide full details of the structural steel section he plans to employ, including calculations showing how the structural steel section fulfils the requirements of the Particular Specification.   The structural steel section shall be in accordance with the relevant British Standard.

## 16.4.  Boring
### 16.4.1.  Boring near recently cast piles

Piles shall not be bored so close to other piles which have recently been cast and which contain workable or partially set concrete that a flow of concrete or instability could be induced or damage caused to any installed piles.   The Contractor's sequence of construction shall be submitted prior to work commencing, in accordance with Section 11 of this Specification.

*Specification for piling and embedded retaining walls.* Thomas Telford, London, 1996.

### 16.4.2. Stability of bore

Temporary casings shall be used in unstable soils with or without support fluid to maintain the stability of the bore unless continuous flight augers are used. The process of advancing the bore and the temporary casing shall be such that soil is not drawn into the bore from outside the area of the pile and cavities are not created outside the temporary casing.

The method of construction shall be submitted in accordance with Section 11 of this Specification.

### 16.4.3. Cleanliness of pile base

Prior to placing steel or concrete, the Contractor shall remove from the base of the pile as much remoulded or loose disturbed materials as practical and in accordance with the method of construction.

### 16.4.4. Cleaning of support fluid

Prior to placing steel or concrete, any support fluid including water shall be wholly or partly removed and replaced while maintaining the fluid head if it does not comply with the Contractor's stated limits for support fluid prior to concreting.

### 16.5. Placement of King Post members

The method of handling and placing King Post members shall be set out in the Contractor's method of construction, submitted in accordance with Section 11.

Unless otherwise specified in the Particular Specification, the member may be plunged into concrete already placed in the pile bore immediately after the concrete placement and before the concrete has achieved its initial set. Guides or hole spacers shall be used so that the specified tolerances of the King Post member are achieved.

Where the King Post member is installed prior to concreting, the Contractor shall ensure that it is not disturbed, displaced or disoriented during concreting.

### 16.6. Steel reinforcement

The number of joints in longitudinal steel bars shall be kept to a minimum. Joints in reinforcement shall be such that the full strength of the bar is effective across the joint and shall be made so that there is no relative displacement during the construction of the pile, following the guidance of BS 8110.

Reinforcement shall be maintained in its correct position during concreting of the wall element. Where it is made up into cages, they shall be sufficiently rigid to enable them to be handled, placed and concreted without damage. If the cage is to be welded together, welding shall be carried out to the requirements of BS 7123. Details of the procedures shall be submitted prior to commencing the Works.

Spacers shall be designed and manufactured using durable materials which shall not lead to corrosion of the reinforcement or spalling of the concrete cover. Details of the means by which the Contractor plans to ensure the correct cover to and position of the reinforcement shall be submitted prior to commencing the Works.

The minimum projecting bond lengths required by the Particular Specification shall be maintained.

### 16.7. Placing and compacting concrete
### 16.7.1. General

The workability and method of placing of the concrete shall be such that a continuous monolithic concrete shaft of the full cross-section is formed, and that the concrete in its final position is dense and homogeneous. Concrete shall be transported from the mixer to the position of the pile in such a manner that it does not cause segregation of the mix. The method of placing shall avoid disturbance of King Post members.

Before commencement of concreting a pile the Contractor shall satisfy himself that the supplier will have available a sufficient quantity of concrete to construct the pile in one continuous operation.

Each batch of concrete in a pile shall be placed before the previous batch has achieved a stiffness which prevents proper amalgamation of the two concrete batches. Removal of temporary casings, where used, shall be completed before the concrete within the casing loses its workability.

No spoil, liquid or other foreign matter shall be allowed to contaminate the concrete.

### 16.7.2. Workability of concrete

The concrete workability shall be determined using the slump or flow table in accordance with BS 1881 and as specified in the Particular Specification. The slump or flow shall be measured at the time of discharge into the pile boring and shall be in accordance with the limits shown in Table 3.1.

### 16.7.3. Compaction

Internal vibrators shall not be used to compact concrete.

### 16.7.4. Placing concrete in dry bores

The concrete shall be poured through a hopper attached to a length of tremie pipe and in such a way that the flow is directed and does not hit the King Post member, the reinforcing bars or the side of the bore. The tremie pipe shall be least 3 m long.

### 16.7.5. Placing concrete under water or support fluid

Concrete to be placed under water or support fluid shall be placed through a tremie pipe in one continuous operation.

The hopper and pipe of the tremie shall be clean and watertight throughout. The pipe shall extend to the base of the boring and a sliding plug or barrier shall be placed in the pipe to prevent direct contact between the first charge of concrete in the pipe of the tremie and the water or support fluid. The pipe shall at all times penetrate the concrete which has previously been placed with a minimum penetration of 3 m and shall not be withdrawn from the concrete until completion of concreting.

At all times a sufficient quantity of concrete shall be maintained within the tremie pipe to ensure that the pressure from it exceeds that from the water or support fluid and workable concrete above the tremie base. The internal diameter of the pipe of the tremie shall be of sufficient size to ensure the easy flow of concrete. It shall be so designed that external projections are minimized, allowing the tremie to pass into the bore without causing damage. The internal face of the pipe of the tremie shall be free from projections.

Where the King Post member is placed in the pile bore prior to concreting, a hopper with twin pipes shall be used and concrete will be placed in such a manner that a balanced concrete level is maintained on both sides of the member.

The depths of the surface of the concrete shall be measured and the embedded length of the tremie pipe recorded at regular intervals corresponding to the placing of each load of concrete. The depth measured and volumes of concrete placed shall be plotted immediately on a graph during the concreting process and compared with the theoretical relationship of depth against volume.

### 16.7.6. Time period for excavation and placing concrete

The time period after a pile is excavated and before the concrete is placed shall not exceed 12 hours unless otherwise specified in the Particular Specification.

If temporary casing which has the same diameter as the pile bore is used, this period shall start when excavation below the temporary casing commences.

### 16.8. Tolerances

The required tolerances depend on the application and are set out in the Particular Specification.

### 16.9. Concrete level

No pile shall be cast with its final concreted level below standing water level unless the Contractor's method of construction includes measures to prevent inflow of water that may cause segregation of the concrete as the temporary casing is withdrawn. The standing water level will be treated as the cut-off level for the purpose of calculating tolerance.

For piles cast in dry bores using temporary casing and for piles cast under water or support fluid the casting level above cut off level shall be within the casing tolerance above the cut off level shown in Table 3.2.

### 16.10. Backfilling

The pile bore containing the King Post member shall be backfilled once the concrete has sufficient strength or after such time as specified in the Particular Specification.

The backfill material in accordance with the Particular Specification shall be evenly placed on either side of the King Post member to avoid uneven loading. Prior to backfilling, bores shall be clearly marked and fenced off so as not to cause a safety hazard.

### 16.11. Instrumentation

Movement of the wall, the surrounding ground and existing structures shall be monitored in accordance with the Particular Specification and Section 19 of this Specification.

# 17. Steel sheet piles

## 17.1. General

All materials and work shall be in accordance with Sections 11 and 17 of this Specification, except where there may be conflict of requirements, in which case this Section shall take precedence.

## 17.2. Particular Specification

The following matters are, where appropriate, described in the Particular Specification:

(a) performance criteria for movement under lateral loads
(b) grades of steel
(c) minimum section modulus, web thickness of sheet pile
(d) surface preparation
(e) types of coating
(f) thickness of primer and coats
(g) types of head and toe preparation
(h) minimum length of sheet pile to be supplied
(i) water retention
(j) restriction on working hours during which driving can take place
(k) types of pile shoe
(l) penetration or depth or toe level
(m) driving resistance or dynamic evaluation or set
(n) Preboring and jetting or other means of easing pile drivability
(o) detailed requirements for driving records
(p) welding of clutches
(q) requirement to carry vertical load
(r) other particular requirements.

## 17.3. Ordering of piles

The Contractor shall ensure that the piles are available at the time for incorporation in the Works. All piles and production facilities shall be made available for inspection at any time. Only new piles shall be used for permanent works. Piles shall be carefully examined at the time of delivery and damaged piles repaired or replaced. The records of testing of the steel used in the piles shall be submitted prior to commencing the Works.

## 17.4. Materials
### 17.4.1. Standard sheet piles

Unless specified otherwise, all steel for sheet piles shall be manufactured to BS 4360, Grades 43A or 50A or to BS EN10 025 Grades Fe 430A or Fe 510A.

The dimensional tolerances of the sheet piles shall comply with Table 17.1.

### 17.4.2. Fabricated sheet piles

All fabricated piles, e.g. corners, junctions, box sections, high modulus sections, shall be fabricated and supplied to the sheet pile manufacturer's recommendations.

### 17.4.3. Storage

If sheet piles of different grade steel are stored on site, each pile shall be clearly marked as to its grade and piles of different grade shall be stored separately.

### 17.4.4. Clutch sealant

If specified in the Particular Specification, the Contractor shall apply a clutch sealant to the piles prior to pitching in accordance

Table 17.1. *Dimensional tolerances for steel sheet piles*

| Dimension to which tolerance applies | | Tolerance | |
|---|---|---|---|
| | | Z type | U type |
| 1. | Width<br>(a) Single piles<br>(b) Interlocked piles | ±2%<br>±3% | ±2%<br>±3% |
| 2. | Thickness of section | +10%, −8% | ±5% |
| 3. | Weight | ±5% | ±5% |
| 4. | Length | ±200 mm | ±200 mm |
| 5. | Squareness of cut for each<br>   section<br>(a) parallel to line of wall<br>(b) perpendicular to line of wall | ≤2%<br>≤10 mm | ≤2%<br>≤10 mm |
| 6. | Straightness | ≤ 0.2% of pile length | ≤ 0.2% of pile length |
| 7. | Depth of section | ±4% | ±4% |

with the manufacturer's recommendations. He shall supply details with his tender of the brand and properties of clutch sealant he plans to use.

## 17.5. Pile handling and driving
### 17.5.1. Pile handling

When assembling piles before pitching the Contractor shall ensure that the interlocks are clean and free from distortion. All storage, handling, transporting and pitching of piles shall be carried out in such a way that no significant damage occurs to piles and their coatings.

### 17.5.2. Pile installation

The Contractor shall satisfy himself that the sheet piles can be installed adequately to the correct depths through the reported or anticipated soil conditions. The Engineer shall be notified 24 hours before the commencement of driving.

The piles shall be guided and held in position by temporary gates with each pile properly interlocked with its neighbour. Piles shall not by-pass one another in place of interlocking.

Where sheet piles are driven in panels, the end piles to each panel shall be driven in advance of the general run of piles. After allowing for initial penetration, no pile in the panel shall be driven to an excessive lead in comparison with the toe level of the panel in general and where hard driving is encountered this lead should not exceed 1 m.

At all stages during driving the free length of the sheet pile shall be adequately supported and restrained. The Contractor shall ensure that the sheet panels are driven without significant damage or declutching.

The selection of driving and contracting plant shall be made having due regard to the ground conditions and pile type. The sheet piles shall be driven to the specified level and/or resistance or if hard driving is experienced, to practical refusal (which shall be defined as when the rate of penetration is below 100 mm per minute when hammering continuously or 12 blows per 25 mm movement when using the appropriate equipment), or when visible damage to the pile occurs. Practical refusal for pile extraction shall be defined generally as when the rate of extraction of a pile is below 100 mm per minute when back-hammering or pulling (with equipment normally capable of withdrawing a pile) continuously (after an initial effort of 10 minutes), or when damage to the pile head occurs. If the piles have not penetrated to the levels specified in the Particular Specification, or have encountered obstructions, the Contractor shall submit details as to how he will overcome the problem.

If required in the Particular Specification, the Contractor shall install the sheet piles using a vibrationless jacking system.

Pile driving hammers shall be correctly positioned on the pile so that the hammer will be aligned as near to the axis of the pile as is practically possible. Freely suspended piling hammers shall be equipped with correctly adjusted leg guides and inserts. Where a hammer is mounted in a rigid leader, the leader shall be stable. The anvil block or driving plate shall be of sufficient size to cover as much as possible of the cross-section of the pile.

Piles previously driven shall not be used until the Contractor can demonstrate that they can meet all the requirements of the Specification.

The Contractor may provide each pile in more than one length. Spliced joints shall be designed to cater for the combined effects of bearing, shear and bending stresses imposed upon the sheet piles. Splices shall be located to avoid maximum stress positions. If splices are to be welded, then these shall be designed in accordance with the guidelines given in BS 5135 and the manufacturer's recommendations. Weld metal shall not encroach within the interlock areas so as not to interfere with the interlocking of the piles.

## 17.6. Positional and alignment tolerance

Unless deflected by obstructions, sheet piles shall be installed within the following tolerances:

- In plan ±75 mm of the given sheet pile line at commencing surface
- Vertical 1 in 75
- Level ±50 mm of required top level.

Pile line dimensions shall be based on the nominal size of piles. Creep or shrinkage of the pile lines shall not exceed the manufacturer's rolling tolerances.

## 17.7. Welding
### 17.7.1. Welders' qualifications

Only welders who are qualified to BS EN 287 and have a proven record over the previous six months, or who have attained a similar standard, shall be employed on the Works. Proof of welders' proficiency shall be made available to the Engineer on request.

### 17.7.2. Welding standard

For manual metal arc and semi-automatic welding of carbon and carbon manganese steels, welding of piles and steel framework shall be carried out in accordance with BS 5135, the standard being Quality Category D in accordance with Appendix H, Tables 18 and 19. Defective welds shall be cut out and replaced. Where steel piles are to be spliced by butt welding the interlocks shall not be welded unless a sealing weld is required.

## 17.8. Durability and protection

Protective coating shall be applied, if specified, following the procedures set out in Section 6 of this Specification.

If a structure is to be welded to piles, the piles shall be cut square and to within ±5 mm of the levels specified. If pile heads are to be encased in concrete they shall be cut to within ±20 mm of the levels specified, and protective coatings shall be removed from the surfaces of the pile heads down to a level 100 mm above the soffit of the concrete.

### 17.9. Piling records

The following records shall be kept where appropriate:

(a) pile reference number or location
(b) pile type and grade of steel
(c) pile length
(d) type of hammer
(e) date of driving
(f) commencing surface level
(g) depth driven
(h) length of offcuts
(i) length of pile extensions
(j) if required, the measurement of driving resistance at appropriate depths
(k) all information regarding interruptions, unexpected changes in driving characteristics, obstructions and times taken in overcoming them.

# 18. Integrity testing of wall elements

**18.1. Method of testing**

Where integrity-testing of wall elements is specified the sonic logging method normally shall be used. Other methods may be considered by the Engineer subject to satisfactory evidence of performance.

**18.2. Particular Specification**

The following matters are, where appropriate, described in the Particular Specification:

    (a)   the method of test to be carried out

    (b)   the number and location of wall elements to be tested

    (c)   the stages in the programme of works when a phase of integrity testing is to be carried out

    (d)   the number and location of wall elements in which ducts are to be placed and number and length of ducts to be provided in each wall element for the sonic logging method

    (e)   the time after testing at which the test results and findings shall be available to the Engineer, if different from the requirements of Sub-section 18.7

    (f)   the depth of wall element over which the testing is required, the depth intervals to be not greater than 0.25 m

    (g)   number of days to elapse between wall element casting and integrity testing

    (h)   other particular requirements.

**18.3. Age of wall elements at time of testing**

Integrity tests shall not be carried out until the number of days specified in the Particular Specification have elapsed since casting of the wall element.

**18.4. Preparation of heads or tops of wall elements**

Where the method of testing requires the positioning of sensing equipment on the wall element head or top, it shall be broken down to expose sound concrete and shall be clean, free from water, laitance, loose concrete, overspilled concrete and blinding concrete and shall be readily accessible for the purpose of testing.

**18.5. Specialist sub-contractor**

The testing shall be carried out by a specialist firm, subject to demonstration to the Engineer of satisfactory performance on other similar contracts before the commencement of testing.

    The Contractor shall submit to the Engineer the name of the specialist integrity testing firm, a description of the test equipment, a test method statement and a programme for executing the specified tests prior to commencement of the Works.

**18.6. Interpretation of tests**

The interpretation of tests shall be carried out by competent and experienced persons.

    The Contractor shall give all available details of the ground conditions, element dimensions and construction method to the specialist firm before the commencement of testing in order to facilitate interpretation of the tests.

## 18.7. Report

Preliminary results of the tests shall be submitted to the Engineer within 24 hours of carrying out the tests.

The test results and findings shall be given to the Engineer within 10 days of the completion of each phase of testing.

The report shall contain a summary of the method of interpretation including all assumptions, calibrations, corrections, algorithms and derivations used in the analyses. If the results are presented in a graphical form, the same scales shall be used consistently throughout the report. The units on all scales shall be clearly marked.

## 18.8. Anomalous results

In the event that any anomaly in the acoustic signal is found in the results indicating a possible defect in the wall element the Contractor shall report such anomalies to the Engineer immediately. The Contactor shall demonstrate to the Engineer that the wall element is satisfactory for its intended use or shall carry out remedial works to make it so. Sonic logging tubes shall be grouted up after the Contractor has demonstrated that the wall element is satisfactory.

# 19. Instrumentation for piles and embedded walls

## 19.1. Particular Specification

The following matters are, where appropriate, described in the Particular Specification:

(a) the type of instrumentation required

(b) the pile or wall element numbers and locations in which the instrumentation is to be installed

(c) the depth or location within the pile or wall where the instrumentation is to be installed

(d) time at which the base readings should be taken

(e) time interval between readings

(f) monitoring equipment

(g) expected load, pressure, displacement or strain range for which results are required

(h) type of loading, compressive or tensile

(i) type of output required

(j) whether remote or direct reading is required

(k) whether the instrumentation monitoring equipment will become the property of the Employer

(l) whether surveying of the pile or element head terminal is required, to what grid and datum and frequency of surveying

(m) responsibility for instrumentation, monitoring and interpretation of results

(n) aims and objectives of instrumentation

(o) other particular requirements.

## 19.2. Type of instrumentation

Where the installation of instrumentation is called for, the type of instrumentation shall be one of the following, as specified in the Particular Specification:

(a) extensometer (rod or magnetic)

(b) inclinometer

(c) load cell

(d) pressure cell

(e) strain gauge (to be attached to steel or precast concrete or embedded in cast-in-situ concrete).

Other methods may be considered by the Engineer subject to satisfactory evidence of performance. All equipment used shall be suitable for its specified purpose.

The instrumentation shall be robust and shall be a proprietary system supplied by a reputable supplier and shall be installed by a Specialist Instrumentation Contractor. The Contractor shall submit details of the Supplier, the instrumentation and curricula vitae for the staff who will install the instrumentation on site, monitor it and analyse the readings. The Contractor shall also submit details of projects where the Specialist Instrumentation Contractor has successfully installed and monitored the specified type of instrumentation.

*Specification for piling and embedded retaining walls.* Thomas Telford, London, 1996.

## 19.3. Extensometers
### 19.3.1. General

The instrumentation shall be securely attached to the reinforcement cage so that no component is displaced during placing of the reinforcement cage or concreting. During concreting, the tubing shall be adequately covered at both ends to prevent the ingress of concrete and in the case of a magnetic extensometer it shall then be filled with clean water. Compression or extension couplings shall be installed as necessary for the movement range specified in the Particular Specification.

## 19.3.2. Rod extensometers

The system will be such that the anchor and reference blocks shall be securely cast into the pile concrete. The rod shall be made of fibreglass or similar material so that it will not corrode or distort nor change its length due to heat or water changes.

The top of the rod shall incorporate a range adjuster with a travel of 25 mm.

If direct measurements are specified the displacement measuring device shall consist of a rechargeable digital dial gauge with a 240 v mains battery charger or analogue dial gauge. The orientation of the dial gauge relative to the top of the tube shall be constant for every reading. If remote reading is specified the readings shall be made by a linear potentiometer or other suitable device, securely held in place.

## 19.3.3. Magnetic extensometers

The system shall be so as that the magnets are securely cast into the pile concrete. The diameter of the tubes and magnets used shall be of the rock extensometer type. The internal diameter of the tube shall be compatible with the proposed measuring device to avoid variations in readings due to non-concentric position of the reading device in the tube. The tube shall be kept full of water during monitoring and the temperature of the water shall be monitored and recorded to an accuracy of 1°C.

One of the following devices shall be used for measuring depths according to the accuracy required:

- Reed switch mounted on a hand held tape similar to that used for ground instrumentation. Readings shall be obtained by sighting a precise level on to the tape at the pile head to achieve a repeatable accuracy of better than 0.3 mm. The absolute level of the precise level shall be monitored during the test; or
- Reed switch mounted on a cable and attached to a micrometer device at the head of the pile such that readings can be established to an accuracy of better than 0.1 mm. The absolute level of the precise level shall be monitored during the test.

## 19.4. Inclinometers

The access tubing shall be securely attached to the reinforcement cage so that no component is displaced during placing of the reinforcement cage or concreting. During concreting, the tubing shall be adequately covered at both ends to prevent the ingress of concrete. Alternatively, nominal 100 mm internal diameter steel duct sealed at the lower end and fitted with a removable screw cap at the upper end, shall be attached to the reinforcement cage. The inclinometer access tube can then be installed afterwards using a cementitious grout containing a non-shrink additive.

The system shall consist of 60 mm nominal diameter corrosion-proof access tubing which shall have four longitudinal internal keyways on two orthogonal axes. The keyways shall be continuous over the length of the access tubing so that the wheels of the

*Specification for piling and embedded retaining walls.* Thomas Telford, London, 1996.

101

torpedoes can pass freely along it. For embedded walls, the keyways shall be oriented so that one set are parallel to the face of the wall. The coupling between sections of the access tube should ensure that the correct alignment of keyways is maintained through a joint. Jointing rivets to connect the tube sections should be located midway between keyways. Sealing mastic should be applied to each rivet head and to the points where the tubing enters a coupling. All joints should be sealed by liberally wrapping with a fabric tape impregnated with a waterproofing compound. This procedure should also be carried out for the end-caps.

The inclinometer readout equipment shall comprise a biaxial torpedo complete with operating cable, cable reel, grip, carrying case, cartridge readout unit or data logger with internal solid state memory and automatic battery charger. The biaxial torpedo shall be fully waterproof, stainless steel and incorporate two robust force servo accelerators aligned at right angles to each other. The torpedo shall be fitted with two orthogonal pairs of centre sprung wheels and shall have a gauge length of 0.5 m and be capable of negotiating a tube curvature of 3 m. The cable should be of an appropriate length. The cartridge readout unit or data logger shall have an alphanumeric LCD and rechargeable battery with sufficient power for 20 hours' continuous use. The memory will be sufficient to accommodate 20 000 readings with time and date. The Contractor shall supply a calibration frame for checking the readout equipment.

Alternatively a string of electrolevels may be used. Sufficient electrolevels shall be incorporated to provide an accurate deformation profile.

The 'azimuth deviation' shall be recorded for each inclinometer access tube and shall be taken into account when processing the data. Where azimuth correction is necessary, the Contractor shall submit details of the method used.

Readings shall be taken for the full length of each inclinometer access tube for faces A, B, C and D. The top wheel of the torpedo will travel up face A first followed by faces B, C and D.

If a full set of readings cannot be obtained due to accumulation of debris in the access tubes, the Contractor shall flush the tubes with water to remove the debris until readings can be taken.

The readings shall be taken in increments of 0.5 m starting from the base of the access tube. All readings shall be recorded and reported to the nearest 0.1 mm. The readings are considered to be sufficiently accurate only if the 'face errors' in the A/B plane and the C/D plane are less than 1.5 mm.

The datum readings to be used in the calculations for each inclinometer shall be notified to the Engineer.

The data shall be processed assuming a fixed base. The surveyed cumulative horizontal movement of the top of the access tube shall be reported to an accuracy of 0.5 mm. The data shall also be processed assuming a specified offset at the top of the access tube at the request of the Engineer.

The processed data for faces A, B, C and D shall be tabulated to show the following:

(a) deviations and face errors
(b) mean deviation
(c) change in mean deviation
(d) cumulative mean deviation
(e) displacement profiles.

Items (*c*) and (*e*) should also be presented graphically to show the deviation or displacement plotted against depth.

### 19.5. Load cells

The loads to be measured may be either compressive or tensile, as specified in the Particular Specification. The load cells shall have a response time of 2 seconds or less in response to monotonically increasing or decreasing loading.

The monitoring system shall provide a stable signal and any temperature or cable length or other effects on the load signal shall be included in the permissible tolerance. If a hydraulic or pneumatic system is used it shall be rated for and have been tested to twice the anticipated maximum pressure. The load cells and cabling shall be compatible with their intended position within a pile and shall be unaffected by the presence of water or fluid concrete. They shall be sufficiently strong to withstand additional loading due to placing of the reinforcement cage and concreting.

As the cell will give a direct reading of load, it should be positioned so that all the load passes through it without any eccentricity. The loading system shall be safe and stable.

### 19.6. Pressure cells

The pressure cells shall have a response time of 2 seconds or less in response to monotonically increasing or decreasing loading.

The monitoring system shall provide a stable signal and any temperature or cable length or other effects on the pressure signal shall be included in the permissible tolerance. If a hydraulic, oil, pneumatic or mercury-based system is used it shall be rated for and have been tested to twice the anticipated maximum pressure. The pressure cells and cabling shall be compatible with their intended position within a pile and shall be unaffected by the presence of water or fluid concrete. They shall be sufficiently strong to withstand additional loading due to placing of the reinforcement cage and concreting.

If the pressure cell is to be placed against hardened concrete on one or both sides, it shall be provided with a mercury filled re-pressurizing tube to ensure that no gap occurs between the hardened concrete and the pressure cell.

### 19.7. Strain gauges
### 19.7.1. General

The strain gauges shall have a response time of 10 seconds or less. The monitoring systems shall provide a stable signal and any temperature or cable length or other effects on the signal shall be included in the permissible tolerance. The strain gauges and cabling shall be compatible with their intended position within a pile and shall be unaffected by the presence of water or fluid concrete. They shall be sufficiently strong to withstand additional loading due to placing of the reinforcing cage and concreting or pile driving.

### 19.7.2. Strain gauges attached to steel or precast concrete piles

The strain gauges shall be securely attached to a mounting plate welded to the steel for steel piles, or securely bolted to the reinforced concrete for precast concrete piles. The welding procedures of Section 6 shall be followed. The method of attachment shall be sufficiently strong so the gauge is not displaced, nor the wires or cables damaged by driving.

### 19.7.3. Strain gauges embedded in concrete

Strain gauges for cast-in-situ piles shall be either the embedment variety or shall be securely attached to mounting plates welded to 'sister' bars which shall be attached to the reinforcement cage. The methods of attachment shall be such that the gauge is not

*Specification for piling and embedded retaining walls.* Thomas Telford, London, 1996.

103

displaced, nor the wires or cables damaged by placing of the reinforcement, concreting, driving or other processes.

The attachment of sister bars shall be at two locations only, so that the amount of bending induced in the bars and measured by the strain gauges changes by less than $50\mu\varepsilon$.

## 19.8. Readings
### 19.8.1. General

Readings shall be reported in terms of the basic property measured (e.g. volts, hertz, mm of Mercury) and shall be converted to SI or derivative units by means of a calibration constant. The calibration constant and the range over which it is applicable shall be clearly stated.

Readings shall be taken at the times or time intervals specified in the Particular Specification.

### 19.8.2. Calibration and data checking

The instrumentation shall be calibrated prior to incorporation into the Works and a certificate of calibration shall be submitted to the Engineer. For strain gauges a calibration constant can be provided if evidence is available showing that the variation in calibration constant will not vary maximum readings outside the specified tolerance.

The instrumentation shall be calibrated so that its behaviour has been monitored over the range specified in the Particular Specification. The Contractor shall demonstrate that the type of instrumentation can provide a stable, reproducible and repeatable calibration.

All data shall be checked by the Contractor for errors prior to submission. If erroneous data are discovered (e.g. face errors for inclinometer readings greater than 1.5 mm), the Contractor shall take a second set of readings immediately. If the errors are repeated, the Contractor shall determine the cause of the error. Both sets of readings shall be processed and submitted, together with the reasons for the errors and details of remedial works. The Contractor shall rectify any faults found in the instrumentation system for the duration of the specified monitoring period.

All computer data files and calculation sheets used in processing the data shall be preserved until the end of the contract. They shall be made available for inspection at the request of the Engineer.

### 19.8.3. Tolerances

For load or pressure, a tolerance of 1% of the range specified in the Particular Specification is permissible. For strain, a tolerance of $2 \times 10^{-6}$ is permissible. For displacement, a tolerance of 0.1 mm is permissible unless otherwise specified.

Except where explicitly specified in the Particular Specification, no instrumentation device, tubing or cable shall be placed in the concrete cover zone.

### 19.8.4. Report

The results shall be given to the Engineer within 5 working days of the completion of each phase of testing.

The report shall contain the following:

(a) the date and time of each reading
(b) the weather
(c) the name of the person who made the reading on site and the name of the person who analysed the readings together with their company affiliations
(d) the pile or wall element reference number and the depth and identity number of the instrumentation
(e) any damage to the instrumentation or difficulties in reading

(f)    the condition of the pile (e.g. if the reading is being made during a load test, the stage of the test; if a wall is being excavated alongside, the depth of the excavation, etc.)

(g)    the calibration constants or equations that are being applied and the dates they were determined

(h)    a table comparing the specified results with any previous readings and with the base readings

(i)    a graph showing variation of load or pressure or vertical movement or strain with time or horizontal movement with depth. Key dates should be marked with a brief explanation of their significance.

Columns of numbers should be clearly labelled together with units. Numbers should not be reported to a greater accuracy than is appropriate. Graph axes should be linear and clearly labelled together with units.

### 19.8.5. Specialist instrumentation contractor

The installation, monitoring and analysis shall be carried out by a specialist firm, subject to demonstration to the Engineer of satisfactory performance before the commencement of monitoring.

The staff carrying out the monitoring and interpretation of the results shall be competent and experienced with the type of instrumentation used.

### 19.8.6. Monitoring equipment

The monitoring equipment shall become the property of the Employer, if so specified in the Particular Specification. Monitoring shall be carried out either directly at the pile head, or remotely from a monitoring cabin as specified in the Particular Specification.

The monitoring equipment shall be appropriate for the situation in which it is to be used. The manufacturer's guidelines for its use shall be followed. Where it has not been possible to follow the manufacturer's guidelines, the fact shall be reported together with the reason and details of the alternative procedure.

The monitoring equipment shall remain on site for the duration of the monitoring programme, except when necessary for it to be calibrated, after which it shall be returned to site. The same monitoring equipment shall be used at each position. Where this is not possible the fact shall be reported together with the reason.

### 19.9. Protection

Terminal boxes at the head of the pile shall be protected by a lockable robust steel cap. The pile head shall then be fenced off with clearly visible barriers which shall be maintained for the duration of the monitoring programme.

Any cable running along the ground shall be clearly marked and adequately protected to prevent the cable posing a safety hazard or becoming damaged.

### 19.10. Surveying

The pile or element head terminal (top of access tubing or as specified in the Particular Specification) shall be surveyed, if specified in the Particular Specification. The level will be determined to an accuracy of 1 mm relative to the datum specified in the Particular Specification. The grid coordinates will be determined to an accuracy of 2 mm relative to the grid specified in the Particular Specification. Other measurements shall be to the specified accuracy.

# 20. General requirements for reinforced concrete

**20.1. General**

All materials shall be in accordance with Section 1 or Section 11 as appropriate of this Specification, the Particular Specification and this Section, except where there may be conflict of requirements, in which case those in the Particular Specification, BS 5328 and this Section shall take precedence.

**20.2. Cement**
*20.2.1. Type of cement*

Cement shall be ordinary Portland, sulphate-resisting Portland, Portland blast-furnace cement or rapid-hardening Portland cement unless otherwise specified. When forwarding his piling method statement and programme to the Engineer, the Contractor shall submit details of the type of cement, other than ordinary Portland cement, he proposes to use. High alumina cement shall not be used.

Cement shall be used in the order it is received on site.

All Portland cements used in the Works shall be obtained from a registered firm under the BSI Scheme for Quality Assurance to BS 5750.

When directed the Contractor shall obtain Certificates of Conformity in accordance with Appendix A of the Quality Assessment Schedule to BS 5750.

*20.2.2. Cement replacement materials*

The use of cement replacement materials will be permitted provided that they can be shown to have no deleterious effects.

*20.2.3. Storage of cement and cement replacement materials*

All cement and cement replacement materials shall be stored in separate containers according to type in waterproof stores or silos.

**20.3. Aggregate**
*20.3.1. Type of aggregate*

Aggregates shall consist of naturally occurring material unless otherwise specified or ordered. The Contractor shall inform the Engineer of the source of supply of the aggregates before the commencement of work and, at the request of the Engineer, provide evidence regarding their properties and consistency.

Where shell is present in the aggregate, the content shall be limited as shown in Table 20.1.

The total chloride ion content of the mix used in steel-reinforced concrete, whether arising from aggregates, water, admixtures or any other source, shall not exceed 0.4% of the mass of cement used (inclusive of any cement replacement material) where the cement is in accordance with BS 12, BS 146, BS 1370 or BS 4246. Where

*Table 20.1. Limit of shell content of aggregate*

| Nominal maximum size of aggregate | Maximum shell content as calcium carbonate ($CaCO_3$) percentage by weight of dry aggregates |
|---|---|
| Over 10 mm | 8 |
| 10 mm and below | 20 |

sulphate-resisting Portland cement to BS 4027 is used, the permitted chloride ion content shall be reduced to half the amount stated above.

For use in pre-tensioned prestressed concrete work, the chloride ion content shall be less than 0.1% of the mass of cement used. The limit of 0.1% shall also apply to the main concrete of post-tensioned prestressed concrete work unless there is an impermeable and durable barrier, in addition to any grout, between the main concrete and the tendons.

Marine aggregates will not be used in prestressed concrete work.

### 20.3.2. Storage of aggregates

All aggregates brought to the Site shall be free and kept free from deleterious materials that will affect the strength or durability of the concrete. Aggregates of different types and sizes shall be stored separately in different hoppers or different stockpiles.

### 20.4. Water
### 20.4.1. Suitability

Water for the Works shall comply with BS 5328.

### 20.4.2. Tests

When required by the Engineer, the Contractor shall arrange for tests of the water for the Works to be carried out in accordance with BS 3148 before and during the progress of the work.

### 20.5. Admixtures

Admixtures complying with BS 5075 may be used if approved and shall be used as and when required.

No admixtures will be permitted which contain more than the equivalent of 0.02% of anhydrous calcium chloride by weight of the cement in the final mix.

### 20.6. Concrete mixes
### 20.6.1. Grade designation

Grades of concrete shall be denoted by the characteristic 28 day test cube strength in Newtons per square millimetre.

### 20.6.2. Mix

Concrete mixes shall be in accordance with Clause 20.6.3 (designed mix) or Clause 20.6.4 (prescribed mix) and of grades 20, 25, 30, 35, 40, 50 or 60 of BS 5328.

The Contractor shall submit the workability he proposes before work commences. Neither trial mixes nor strength tests are required for prescribed mixes.

The concrete shall have sufficient workability to enable it to be placed and compacted by the methods used in forming the piles.

In order to ensure sufficient resistance to Alkali Silica Reaction (ASR), the constituent materials in concrete mixes shall comply with one or more of the following requirements:

(i) coarse aggregates of granite or limestone in conjunction with fine aggregates all of which are certified by a qualified chemist competent in concrete chemistry as unlikely to be reactive.

(ii) the use of 'low alkali' sulphate resisting cement with an alkali content not greater than 0.6% measured as the equivalent $Na_2O$ (mean acid-soluble alkali-level) as determined by the method of BS 4550: Part 2, clause 16.

(iii) the total equivalent $Na_2O$ (acid-soluble) alkali content from the cement shall not exceed 3 kg per cubic metre of concrete, expressed in the following terms:

| Cement content (kg/m³ concrete) | Maximum alkali content of cement: % |
|---|---|
| 500 | 0.60 |
| 400 | 0.75 |
| 375 | 0.80 |
| 330 | 0.90 |
| 300 | 1.00 |

based on the equation:

$$\text{Maximum \% acid-soluble alkali content of cement} = \frac{300}{(\text{Cement content in kg/m}^3)}$$

(*iv*) cementitious material shall contain, by weight, a minimum of either 50% ground granulated blast furnace slag or 30% pulverized fuel ash.

Evidence of compliance shall be provided before commencing the Works.

The total chloride content of the concrete mix shall not exceed the limits in Table 6.4 of BS 8110. The total chloride content shall be calculated from the mix proportions and measured chloride content of each of the constituents.

The total water soluble sulphate content of the concrete mix, expressed as $SO_3$ shall not exceed 4% by mass of the cement in the mix. The sulphate content shall be calculated as the total from the various constituents of the mix.

### 20.6.3. Designed mix

When a designed mix is specified the Contractor shall be responsible for selecting the mix proportions to achieve the required strength and workability,

Designed mixes shall be in accordance with grades 20, 25, 30, 35 or 40 (for cast-in-place piles) or grades 35, 40, 50 or 60 (for precast piles) of BS 5328. Complete information on the mix and sources of aggregate for each grade of concrete and the water/cement ratio and the proposed degree of workability shall be provided to the Engineer before work commences.

### 20.6.4. Prescribed mix

Prescribed mixes for cast-in-place piling shall be in accordance with Table 20.2.

### 20.6.5. Minimum cement content

The cement content in any mix shall be not less than 300 kg/m³. Where concrete is to be placed under water or drilling mud by

*Table 20.2. Prescribed mixes for cast-in-place piling using nominal maximum aggregate size 20 mm*

| Concrete grade | Piling mix workability (see Table 3.1) | Slump limits, mm, or target flow, mm, if marked with * | Prescribed mix | | | |
|---|---|---|---|---|---|---|
| | | | Cement: kg/m³ | Total aggregate: kg/m³ | Percentage sand in total aggregate | |
| | | | | | Grading C | Grading M |
| 20 | A | 75–150 | 320 | 1840 | 37 | 35 |
| | B | 400–500 * | 350 | 1790 | 40 | 38 |
| | C | 500–600* | 400 | 1740 | 40 | 38 |
| 25 | A | 75–150 | 360 | 1790 | 37 | 35 |
| | B | 400–500 * | 400 | 1740 | 40 | 38 |
| | C | 500–600* | 460 | 1680 | 40 | 38 |
| 30 | A | 75–150 | 410 | 1750 | 37 | 35 |
| | B | 400–500 * | 460 | 1680 | 40 | 38 |
| | C | 500–600* | 520 | 1600 | 40 | 38 |

*Specification for piling and embedded retaining walls.* Thomas Telford, London, 1996.

tremie the cement content shall be not less than $380 \text{ kg/m}^3$, or where the pile will be exposed to sea water $400 \text{ kg/m}^3$.

## 20.7. Trial mixes
### 20.7.1. General

When designed mixes are specified, trial mixes shall be prepared for each grade of concrete in accordance with BS 1881. Sampling and testing of the concrete shall be in accordance with BS 1881.

### 20.7.2. Preliminary trial mixes

When required in accordance with Clause 20.7.1 the Contractor shall, before the commencement of concreting, have preliminary trial mixes prepared, preferably under full-scale production conditions or, if this is not possible, in a NAMAS accredited laboratory using a sufficient number of samples to be representative of the aggregates and cement to be used. The results of trial mixes shall be obtained from three separate batches of concrete made using the proposed mix and constituent materials and under full-scale production conditions.

The workability of each trial batch shall be determined and shall be within the tolerances stated in BS 5328.

Three cubes shall be made from each batch, and shall be tested at 28 days. The average strength of the nine 28 day cubes shall exceed the specified characteristic strength by not less than $11.5 \text{ N/mm}^2$. Alternatively, earlier tests on nine cubes shall demonstrate that the specified characteristic strength will be exceeded by $11.5 \text{ N/mm}^2$.

When accelerated testing is proposed for works cubes, an additional four cubes from each batch shall be made, cured and tested in accordance with the accelerated regime.

### 20.7.3. Trial mixes during the work

Where a trial mix is required after commencement of the work, the procedure in Clause 20.7.2 shall be adopted for full-scale production conditions. The strength requirement shall be as in Clause 20.6.3.

### 20.7.4. Workability

The workability of each batch of the trial mixes shall be determined by the slump test as described in BS 1881 or flow table to BS 1881, Part 105, and as specified in the Particular Specification.

### 20.7.5. Variations in mix

No variations outside the limits set out in BS 5328 shall be made in the proportions, the original source of the cement and aggregates, or their type, size or grading zone without demonstrating compliance with this Specification.

## 20.8. Testing works concrete
### 20.8.1. Sampling

Concrete for piles shall be sampled in accordance with BS 5328. Samples shall be taken from the mixer or delivery vehicle at the point of placing the concrete.

### 20.8.2. Workability

The workability of concrete shall be determined by the slump test as described in BS 1881 or flow table to BS 1881, Part 105, and as specified in the Particular Specification. The workability of each batch shall be measured and reported to the Engineer.

### 20.8.3. Works cube tests

For each grade of concrete four cubes shall be made from a single batch for each $15 \text{ m}^3$ of concrete or part thereof in each day's work. The cubes shall be made, cured and tested in accordance with BS 1881. Testing shall be carried out by a NAMAS accredited laboratory. One cube shall be tested at an age of 7 days, two at 28 days, and one cube shall be held in reserve for further testing as required. The Contractor shall submit certified copies of the results of all tests to the Engineer.

*Specification for piling and embedded retaining walls.* Thomas Telford, London, 1996.

109

| | |
|---|---|
| **20.8.4. Standard of acceptance** | The standard of acceptance of the works cubes shall be in accordance with BS 5328. Characteristic strength of concrete shall mean that value of strength below which no more than 5% of the works test results for each concrete grade will fall. |
| **20.8.5. Records of tests** | The Contractor shall keep a detailed record of the results of all tests on concrete and concrete materials. Each test shall be clearly identified with the piles to which it relates. |
| **20.9. Batching concrete**<br>**20.9.1. General** | Facilities shall be provided for the Engineer to inspect the concrete mixing plant or plants when requested.<br>    Unless otherwise specified the requirements in Clauses 20.9.2, 20.9.3 and 20.9.4 shall be met. |
| **20.9.2. Accuracy of weighing and measuring equipment** | The weighing and water-dispensing mechanisms shall be maintained at all times to within the limits of accuracy described in BS 1305. |
| **20.9.3. Tolerance in weights** | The weights of the quantities of each size of aggregate and of cement shall be within 2% of the respective weights per batch after due allowance has been made for the presence of free water in the aggregates, which shall be determined by the Contractor. |
| **20.9.4. Moisture content of aggregates** | The moisture content of aggregates shall be measured immediately before mixing and as frequently thereafter as is necessary to maintain consistency of mix and other specified requirements. |
| **20.10. Mixing concrete**<br>**20.10.1. Type of mixer** | The mixer shall be of the batch type, and shall have been manufactured in accordance with BS 1305 or shown by tests in accordance with BS 3963 to have mixing performance within the limits of Table 5 of BS 1305, 1974. |
| **20.10.2. Tolerance of mixer blades** | The mixing blades of pan mixers shall be maintained within the tolerance specified by the manufacturers of the mixers, and the blades shall be replaced when it is no longer possible to maintain the tolerances by adjustment. |
| **20.10.3. Cleaning of mixers** | Mixers which have been out of use for more than 30 minutes shall be thoroughly cleaned before another batch of concrete is mixed. Unless otherwise specified by the Engineer, the first batch of concrete through a mixer shall contain the normal batch quantity of cement and sand, but only two thirds of the normal quantity of coarse aggregate. Mixing plant shall be cleaned thoroughly between the mixing of different types of cement. |
| **20.10.4. Minimum temperature** | The temperature of fresh concrete shall not be allowed to fall below 5°C. No frozen material or materials containing ice shall be used. In cold weather when the ambient air temperature is less than 5°C, the heads of newly cast piles are to be covered to protect them against freezing unless the final cut-off level is at least 0.25 m below the final head level as cast. Where a pile is cast in frozen ground, appropriate precautions shall be taken to protect any section of the pile in contact with the frozen soil where this occurs below the cut-off level. In the construction of precast piles the requirements of BS 5328 shall be observed. |

## 20.11. Transporting concrete
### 20.11.1. Method of transporting

Concrete shall be transported in uncontaminated watertight containers in such a manner that loss of material and segregation are prevented.

## 20.12. Ready-mixed concrete
### 20.12.1. Conditions of use

Ready-mixed concrete shall be produced at a depot certified as a 'QSRMC accredited plant' under the Quality Scheme for Ready-Mixed Concrete's regulations.

All the constituents for each mix shall be added at the manufacturer's depot. No extra water or other material shall be added after the concrete has left the depot.

Each load shall be accompanied by a delivery note stamped with the time of mixing and stating the consignee and quantities of the constituent materials including water and additions.

### 20.12.2. Mixing plant

Truck mixer units and their mixing and discharge performance shall comply with the requirements of BS 4251.

## 20.13. Steel reinforcement
### 20.13.1. Condition

Steel reinforcement shall be stored in clean conditions. It shall be clean, and free from loose rust and loose mill scale at the time of fixing in position and subsequent concreting.

### 20.13.2. Grade

The grade of steel shall be as specified.

### 20.13.3. Bending of reinforcement

Reinforcement shall be bent in accordance with BS 4466, 1981 to the correct shape before fixing and placing. Samples for bend tests if required will be selected by the Engineer.

### 20.13.4. Placing of reinforcement

Reinforcement in the form of a cage shall be assembled with additional support, such as spreader forks and lacings, necessary to form a cage which can be lifted and placed without permanent distortion. Intersecting bars shall be fixed together. Hoops, links or helical reinforcement shall fit closely around the main longitudinal bars and be bound to them by wire, the ends of which shall be turned into the interior of the pile. Reinforcement shall be placed and maintained in position to provide the specified projection of reinforcement above the final cut-off level.

Unless otherwise specified reinforcement shall extend to the bottom of the excavation, or 3 m below bottom of temporary lining tube, whichever is the least.

Spacers shall be designed and manufactured using durable materials which will not lead to corrosion of the reinforcement or spalling of the concrete cover. Details of the means by which the Contractor plans to ensure the correct cover to and position of the reinforcement shall be submitted.

### 20.13.5. Welding of reinforcement

Welded joints and welding procedures shall be carried out in accordance with BS 7123. Cold-worked reinforcement to BS 4461 shall not be welded.

## 20.14. Grout
### 20.14.1. General

(a) Grout shall have a minimum cement content of 380 kg/m$^3$ or 400 kg/m$^3$ in marine conditions.

(b) The design and workability of grout to be used in the formation of piles shall produce a mix which is suitable for pumping.

(c) Grout shall consist only of ordinary Portland cement, water, fine aggregate (if permitted) and permitted admixtures. Admixtures containing chlorides shall comply with Clause 20.3.1.

*Specification for piling and embedded retaining walls.* Thomas Telford, London, 1996.

111

(d) The fine aggregate shall be in accordance with the limits of grading M or C given in Table 5 of BS 882.

(e) Grout shall have a water/cement ratio as low as possible consistent with the necessary workability, and the water cement ratio shall not exceed 0.40 unless a mix containing an expanding agent is used.

(f) Grout shall not bleed in excess of 2% after 3 hours, or a maximum of 4%, when measured at 18°C in a covered glass cylinder approximately 100 mm in diameter with a height of grout of approximately 100 mm, and the water shall be reabsorbed after 24 hours.

### 20.14.2. Testing works grout

Cube strength testing shall be carried out in accordance with BS 1881. A sample shall consist of a set of six 100 mm cubes. Three cubes shall be tested at seven days and the remaining three at 28 days after casting.

### 20.14.3. Batching grout

The weighing and water-dispensing mechanisms shall be maintained at all times to within the limits of accuracy described in BS 1305.

The weights of each size of aggregate and of cement shall be within 2% of the respective weights per batch after due allowance has been made for the presence of free water in the aggregates.

The moisture content of aggregates shall be measured immediately before mixing and as frequently thereafter as is necessary to maintain consistency of mix.

### 20.14.4. Mixing grout

Cement grouts shall be mixed thoroughly to produce a homogeneous mix.

The grout shall be mixed on site and used within 30 minutes. Following mixing the grout shall be passed through a 5 mm aperture sieve.

### 20.14.5. Transporting grout

Grout shall be transported from the mixer to the position of the pile in such a manner that segregation of the mix does not occur.

### 20.14.6. Placing grout in cold weather

Grout shall have a minimum temperature of 5°C when placed. No frozen material or material containing ice shall be used for making grout. All plant and equipment used in the transporting and placing of grout shall be free of ice that could enter the grout.

# 21. Support fluid

## 21.1. General requirements

Where a support fluid is used for maintaining the stability of an excavation the properties and use of the fluid shall be such that the following requirements are achieved:

(*i*) continuous support of the excavation

(*ii*) solid particles are kept in suspension

(*iii*) the fluid can be easily displaced during concreting

(*iv*) the fluid does not coat the reinforcement to such an extent that the bond between the concrete and reinforcement is impaired

(*v*) the fluid shall not cause pollution of the ground and groundwater before, during or after use.

Details of the type of support fluid, manufacturer's certificates for the constituents and mix proportions shall be submitted at the time of tender.

## 21.2. Particular Specification

The following matters are, where appropriate, described in the Particular Specification:

(*a*) minimum material testing requirements

(*b*) environmental restrictions on use

(*c*) other particular requirements.

## 21.3. Evidence of suitability of support fluid

The Contractor shall provide details of the properties and use of the support fluid to demonstrate it will meet the specified requirements. These details shall be submitted at least 21 days prior to the commencement of work and shall include:

(*i*) evidence from previous work with this support fluid and justification for its suitability for these ground conditions and method of construction; particular issues which should be addressed are the types and sources of the support fluid constituents, time of construction of the piles, ambient temperature, soil and groundwater chemistry

(*ii*) results of representative laboratory or field mixing trials with the support fluid to demonstrate compliance with the Specification

(*iii*) details of the tests to be used for monitoring the support fluid during the Works and the compliance values for these tests, presented in the form of Table 21.1.

## 21.4. Materials
### 21.4.1. Water

If water for the Works is not available from a public supply, the Contractor shall use an alternative source that shall comply with the testing below.

When required by the Engineer, the Contractor shall arrange for tests of the water for the Works to be carried out in accordance with the specified schedule before and during the progress of the work. The frequency of testing shall be as stated in the Particular Specification.

---

### 21.4.2. Additives to the water

All solid additives shall be stored in separate waterproof stores with a raised floor or in waterproof silos which shall not allow the material to become contaminated.

Additives shall generally be used in accordance with the manufacturers recommendations unless demonstrated otherwise. Bentonite shall be of a quality that shall accord with Publication 163: *Drilling fluid materials* of the Engineering Equipment and Materials Users Association.

### 21.5. Mixing of support fluid

The constituents of the fluid shall be mixed thoroughly to produce a homogeneous mix. The temperature of the water used in mixing, and of the support fluid at the time of commencing concrete placement shall not be less than 5°C.

### 21.6. Compliance testing of support fluid

The Contractor shall carry out testing of the support fluid in accordance with his regime and this Specification to demonstrate compliance with his limits for each test. The Contractor shall establish a suitably equipped and properly maintained site laboratory for this sole purpose and provide skilled staff and all necessary apparatus to undertake the sampling and testing.

Each batch of freshly prepared or reconditioned slurry shall be proven by sampling and testing to be within the compliance values and the results submitted before the batch is used in excavations.

*Table 21.1. Test and compliance values for support fluid*

| Compliance values measured at 20°C | | | | |
|---|---|---|---|---|
| Property to be measured | Test method and apparatus | Specification for test | Freshly mixed fluid | Sample from bore prior to placing steel and concrete |
| Density | Mud balance | | | |
| Rheological properties — over appropriate range of shear rates and temperature | | | | |
| (a) Plastic viscosity | Fann viscometer | | | |
| (b) Yield stress | Fann viscometer | | | |
| (c) Consistency index, $K$ | Fann viscometer | | | |
| (d) Flow index, $n$ | Fann viscometer | | | |
| (e) Marsh cone viscosity | Marsh cone | | | |
| (f) Gel strength — for appropriate range of times | Fann viscometer | | | |
| Sand content | Sand screen set | | | |
| Fluid loss — for appropriate range of temperatures times and pressures | Fluid loss test | | | |
| Filter cake thickness | Fluid loss test | | | |
| pH | Electrical pH meter to BS 3445, range pH 7 to 14 | | | |

*Specification for piling and embedded retaining walls.* Thomas Telford, London, 1996.

Details of the method, frequency and locations for sampling and testing slurry from the excavations shall be submitted at least 21 days prior to the commencement of work. At least one sample immediately prior to placing steel and concrete shall be taken and tested from the base of the excavation and one from the top.

If tests show the support fluid does not comply with the Specification, it shall be replaced.

## 21.7. Spillage and disposal of support fluid

All reasonable steps shall be taken to prevent the spillage of support fluid on the site in areas outside the immediate vicinity of boring. Discarded fluid shall be removed from the site without undue delay. Any disposal of fluid shall comply with the requirements of current legislation and all relevant authorities.

# Contract documentation and measurement

# 1. Definitions

In both the *Specification for piling and embedded retaining walls* and this document, reference is made only to the 'Engineer' and to the 'Contractor'. Where piling works are executed under a contract which is not governed by ICE conditions or similar, an allowance has been made in the Particular Specifications for Sections 1 and 11 for other bodies to be nominated as Supervising Officer. In such a case, the Particular Specification could state: 'Wherever in this Specification Engineer is written, then Construction Manager (or whatever) should be read in its place'.

'Contractor' means the main contractor appointed by the Employer to undertake the contract works. The piling contractor may be a nominated or domestic sub-contractor to the Contractor or may be appointed by the Employer as Contractor.

Definitions are made in Sections 1 and 11.

Throughout this volume, where 'piles' or 'piling' are referred to, the reference should be taken to include 'wall elements' or 'embedded retaining walls'. This is to assist with the readability of the text.

*Specification for piling and embedded retaining walls*. Thomas Telford, London, 1996.

119

# 2. Appointment of contractors for piling and embedded retaining walls

## 2.1. Procedure prior to drafting tender documents

Following the completion of all necessary site investigation and a decision that piling is required, the Engineer should consider the possible advantages of pre-tender consultation with specialist contractors.

The Engineer should then decide who is to be responsible for selection of type of pile and for pile design. If more than one type is considered suitable, alternative bills of quantities should be drafted.

If the Engineer decides to delegate responsibility for design of the piles to the contractor, he should provide a performance specification. Guidance on the drafting of a performance specification is given in Appendix G. Any design responsibility so imposed on a contractor should be consistent with the terms and conditions of the main contract. Although a contractor may undertake responsibilities *vis-à-vis* the Engineer, these may not relieve the Engineer of his responsibilities under a contract.

## 2.2. Method of invitation of tenders

Contractors may be invited to tender for piling works as a main contract to an Employer either under one of the established forms of contract, or under a contract document drawn up specially for management type contracts.

Sub-contract tenders for piling works may be invited in the following ways.

(*a*) Tenders may be invited from piling contractors for nomination as sub-contractor to a Contractor; in such cases, it is normal for a prime cost item to be included in the main contract bills of quantities. It is necessary to ensure that the piling specification and other relevant items from piling sub-contract documents

*Fig. 1. Possible effect of changes of commencing surface level on the length of temporary casing required for bored piles; consequences may affect access available for the delivery of concrete and may entail changes of cut-off tolerance*

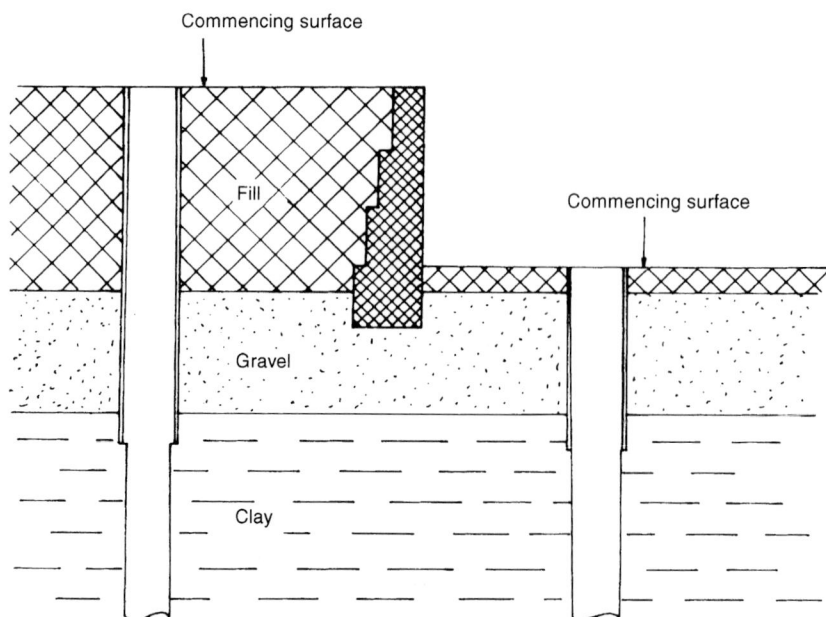

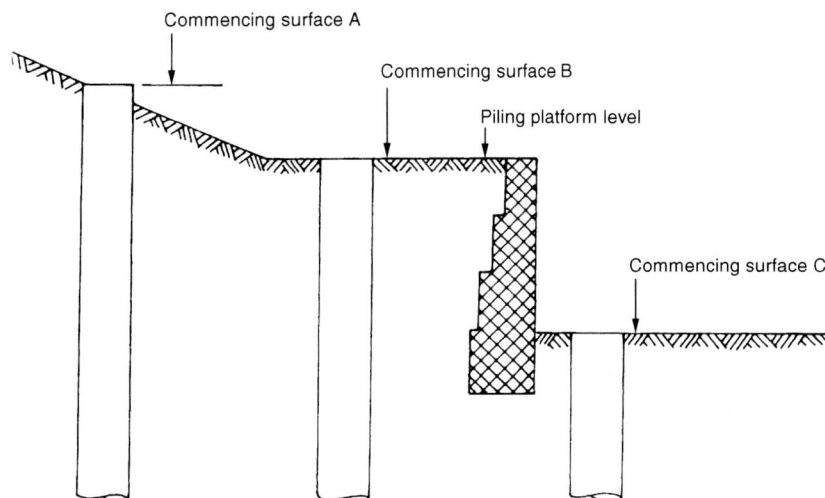

*Fig. 2. Relationship of piling platform level to commencing surface: the piling platform level at which a piling machine sits during pile installation may not correspond with the commencing surface where a pile enters the ground; a piling machine standing on the piling platform level B could install piles with commencing surface higher (A) or lower (C); the relationship of piling platform level and commencing surface level may be significant in the choice of equipment to penetrate to the required depth*

are included in the main contract tender documents so that tenderers for the main contract are aware of and can price for their responsibilities to and attendance on the piling sub-contractor. Equally the sub-contract tender documents should include information regarding the form of the main contract, including the appendix to the contract, in sufficient detail to ensure that a piling sub-contractor is aware of his responsibilities and liabilities at the time of preparing his tender.

(*b*)  The piling works may be measured and included in the bills of quantities for the main contract, and the document may stipulate that the piling works shall be executed by any one firm from a list of approved piling contractors. The Contractor may have the option of proposing further firms for approval.

(*c*)  The piling works may be measured and included in the bills of quantities for the main contract but without any list of approved piling sub-contractors. In such cases, the Contractor will seek prices from his own selection of piling sub-contrac-tors, for whom it is normal practice to seek approval before they are employed on the Works.

(*d*)  Following his appointment, the Contractor may be instructed to carry out additional works not included in the contract. Should this additional work include piling, tenders for such works will normally be invited from a list of firms agreed by the Contractor and the Engineer. It is most desirable that the tender document should be approved by the Engineer before enquiries are issued.

With procedures (*b*), (*c*) and (*d*), the selected piling sub-contractor is normally appointed as a domestic (direct) sub-contractor of the Contractor.

## 2.3. Instructions to tenderers

Although they will not form part of a contract, instructions to tenderers should be provided as necessary and should include

(*a*)  any restrictions relating to visits to the site and the name, address and telephone number of any person from whom permission to visit the site must be obtained, should this be a requirement

(*b*)  any requirement that tenderers submit a programme with their tender

(c) the procedure for submitting a tender and the latest date for its submission

(d) the expected date of award of the contract and the likely commencement date of the works

(e) the name and address of the Contractor, if appointed, in cases where piling works are to be executed as a sub-contract.

## 2.4. Documents forming part of contract

Documents which will form part of a contract (or sub-contract) for piling works are

(a) conditions of contract (and of sub-contract where applicable) and any special conditions required by the Employer

(b) form of tender

(c) general and particular specifications

(d) special clauses appertaining to the piling works (see paragraph 2.5)

(e) drawings scheduled as contract drawings in the tender document

(f) site information and ground investigation data

(g) bill of quantities (see Appendix A)

(h) form of acceptance.

## 2.5. Model clauses suitable for inclusion in contract documents

Model clauses outlining various principles which are appropriate to piling works, whether carried out as a main contract or as a sub-contract, are given in Appendix B.

Where piling is executed under a sub-contract, additional provisions must be incorporated in the documents to define the responsibilities of the Contractor and the sub-contractor. The Federation of Piling Specialists (FPS) have produced a standard list of attendances and facilities which the sub-contractor may require to be provided. This is given in Appendix C. The precise wording of such clauses may need to be varied to suit the particular needs of the contract works and the form of main contract.

## 2.6. Information to be included in tender documents

The following information should be incorporated within the tender documents for piling works, whether as a main contract or as a sub-contract

(a) the period for which the tender is to be valid for acceptance

(b) the location of the site and access thereto

(c) the available working and storage areas

(d) any special conditions limiting noise and vibration

(e) any limitations of working hours

(f) level data, including existing ground levels, commencing surface levels and pile cut-off levels, or, if this information is not available, an indication of the depth of the cut-off levels below the ground surfaces (see Figs 1 and 2); all levels should be related to an Ordnance or other datum as required by the Engineer

(g) any phasing of the works necessitating more than one visit by the (piling) contractor

(h) ground investigation data, including ground levels at borehole and other test positions relative to the stipulated datum (see (f) above) and showing the positions of such boreholes etc. relative to the outline location of the piling works

(i) results of any preliminary pile tests

(j) drawing showing details of underground services, structures, and other known obstructions

(k) details of the allowable pile capacity and specified working load (SWL) for which the piles have been designed, or alternatively the performance specification, including the loads to be supported; if the contractor is required to design and specify the permanent piles, this must be expressly stated in the Particular Specification

(l) if the Engineer responsible for the piling tender documents has not been engaged to supervise the works, it is desirable that this should be so stated.

Additionally, in the case of sub-contract documents, the following should be incorporated

(m) full details of the relevant conditions of the main contract, notably such matters as insurances, retention, liquidated damages, programme requirements, bonds, etc., which may affect the Contractor's tender

(n) any known special conditions/restrictions to be imposed by the intended Contractor.

## 2.7. Information to be supplied by the tenderer

Tender documents should be drawn up so as to give the tenderers the opportunity to provide information required for the evaluation of the tenders. Comments accompanying tenders should not repeat points which are adequately covered in the tender documents. However, in the event that the tender documents do not adequately cover the items set out below, the tenderers should include relevant comments

(a) *validity of tender*: the period for which the tender remains open for acceptance, and the period required for execution of the works

(b) *insurances*: the upper limit to which the tenderer is already insured if it is less than a figure stipulated in the enquiry and will therefore require to be supplemented

(c) *working hours*: the basis of the tender in respect of the days and hours during which the tenderer expects to work, under normal conditions

(d) *method statement*: a description of the type of equipment proposed to be used for the execution of the piling works, the method of construction and, where appropriate, any assumptions made regarding the programme and/or sequence of piling operations, the number of piling machines and number of visits allowed

(e) *contract period*: an estimate of the contract period for working without disruption on work specified at tender stage

(f) *notice*: period of notice required for commencement

(g) *design period*: where applicable, period required for completion of design.

## 2.8. Evaluation of tenders

Tenders for the bulk of civil engineering and building works can be compared equitably by giving primary consideration to the tender sum evaluated from a priced bill of quantities.

Pile lengths required for a given pile capacity in particular ground conditions will not necessarily be the same for piles of different types and sizes. The lowest tender will not necessarily represent best value for money in terms of the final account after remeasurement. When examining tenders for piling works, the Engineer should consider the possible financial effects of variations in the measured quantities relative to each pile type.

## 2.9. Early final payment to piling or walling sub-contractor

Piling works are necessarily carried out in the early stages of main contracts and piling sub-contractors have suffered the inequity of long-term retention pending practical completion of the main contract works. Some current conditions of contract make definite provision for early final payment to nominated sub-contractors, subject to certain safeguards, and such provisions should be incorporated.

*Specification for piling and embedded retaining walls.* Thomas Telford, London, 1996.

# 3. Responsibility for piling and embedded retaining walls

## 3.1. Design

Irrespective of which form of contractual conditions is applicable, the responsibility for design must be stated. Where the Engineer responsible for the design of the overall foundation system is not responsible for the design of the piles, then clear performance requirements must be specified, the piling contractor must be free to vary the pile lengths or diameters to achieve the performance requirements and the contract must include sufficient static pile load testing so that the specified performance can be shown to have been provided.

In some instances the specialist piling contractor is responsible for determining the type of pile to be used to support structural loads as specified. In such cases, he will be responsible for the design of the pile elements to support the structural loads while the Engineer retains full responsibility for the design of the foundation system and associated structure. The Engineer must consider the effects of soil–structure interaction in arriving at the performance requirements for an individual pile under test.

Where the Engineer has designed the piles, the assumptions and criteria used in the design should be stated in the piling tender document. If the Contractor considers the Engineer's design to be inappropriate, he should state so in his tender return.

At the time of drafting these procedural notes, there exists a conflict between the factoring of loads used in limit state codes and the traditional approach to evaluating service loads on piles. Pending the revision and implementation of codes to maintain consistency throughout the design of structures, it is imperative that particular attention be given to ensuring that all loads to be carried by piles are clearly defined (e.g. whether the specified loads are characteristic or ultimate).

## 3.2. Selection of type of pile or embedded retaining wall

It is essential that the type of pile to be specified is carefully considered to ensure its suitability in relation to the ground and environmental conditions. These conditions must be properly defined by means of adequate site and ground investigation works to permit appropriate selection of pile type and the proper design of the pile.

If the type of pile and/or method of construction or of installation is not specified in the tender documents, and is not fully described in the tender submission, then this information should be obtained following receipt of tenders and should be agreed prior to a contract being let.

Criteria governing the length and/or driving resistance of driven piles, or the length of bored piles, testing requirements, etc., should be agreed before entering into a contract.

Decisions affecting the pile design, in consequence of conditions encountered on site, should be the responsibility of the Engineer. In circumstances where significant variations in ground conditions are found, or where it is necessary to found piles deeper than envisaged in differing ground conditions (see Fig. 3), the Engineer may request and consider proposals from the Contractor.

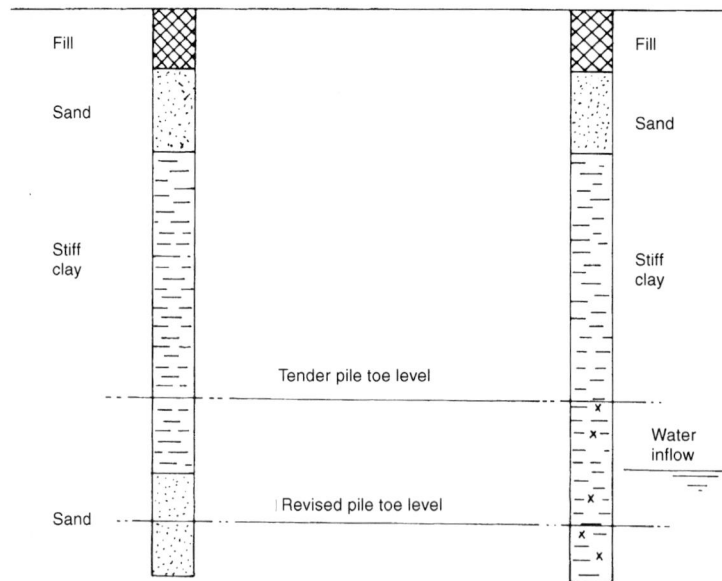

*Fig 3. Possible effects of lowering pile toe levels into differing ground conditions or groundwater regimes: in some such instances it may be necessary to change the type of pile from that envisaged or to place concrete by tremie instead of normally; other consequences may entail a change of the piling equipment due to a revised pile toe level being beyond the reach of the piling plant provided on site or on which a tender has been based; also illustrated is the importance of investigating subsoil conditions to an adequate depth below the pile toe levels originally envisaged; the figure shows only two examples of a wide range of conditions which may significantly affect piles of all types*

## 3.3. Contractual arrangements

Piling may be carried out with the piling contractor acting as the Contractor. More commonly, piling is executed under a nominated or domestic sub-contract to a civil engineering or building Contractor.

There are four basic types of contractual arrangement under which piling may be undertaken.

(*a*)  Civil engineering works with an Engineer responsible to an Employer for design and supervision.

(*b*)  Building works with an Architect responsible to an Employer for design and supervision and advised by an Engineer, who may also be responsible to the Employer for structural engineering elements of the works but who has no formal status under the conditions of contract. Under such a contract, it is desirable that the Architect authorizes the Engineer to act as his representative in connection with piling and other structural engineering elements of the works.

(*c*)  Building or civil engineering works with a Contractor responsible to an Employer for design and construction. The Contractor may appoint an Engineer to undertake the engineering duties appertaining to the piling or may entrust those duties to a suitably qualified engineer on his staff. It is desirable that the Contractor demonstrates to the Employer the devolution of those responsibilities to an Engineer, who should be unfettered in exercising his duties even though the Contractor remains liable for the fulfilment of the contract.

(*d*)  Building works with an Architect responsible to an Employer for design and supervision but having no engineering advisor. The Architect should consider the responsibilities which devolve upon him in these circumstances and should recognize that such an arrangement may not be in the best interests of either his Client or himself.

## 3.4. Information to be provided by the Engineer

The Engineer should ensure that borehole records, test results, details of the groundwater regime and all other relevant factual geotechnical data are incorporated in the tender documents. Contractors tendering for the piling works should not be required, during the tender period, to visit an office to inspect ground investigation reports and similar data: all this information should be included with the tender documents for their careful consideration throughout the tender period.

Adequate and high quality geotechnical information is essential for the proper design and construction of piling works. Reference should be made to the suite of ground investigation documents produced by the Site Investigation Steering Group (Thomas Telford, 1993).

The Engineer should be responsible for advising the Employer on the scope of ground investigation required for the piling works and for the interpretation of the information obtained. It is preferable that the interpretative report should be included in the information provided to the Contractor. However, the provisions of Clause 1.7 of the *Specification for piling and embedded retaining walls* provide for the situation where opinions or conclusions are embodied in such ground investigation data.

Piling works are sometimes presented in a bill of quantities for pricing by firms tendering for a main contract. In such circumstances, the tenderers have to obtain quotations from specialist sub-contractors during the tender period and difficulties can arise when all relevant geotechnical information, piling specifications, relevant preambles, etc., are not readily identifiable without close scrutiny of large bills of quantities. Such information should, wherever possible, be collated in a single section within the bill or otherwise readily identified so that the firms tendering for the main contract can issue readily all relevant information to piling sub-contractors.

The sub-contractor should be provided with the above information with sufficient time to enable him to evaluate the situation and design the piles before submission of his tender.

## 3.5. Workmanship

The Contractor is responsible both for the workmanship for the proposed methods of construction based on the information made available and/or accessible to him.

Proper supervision of piling works by experienced site personnel is essential for the sound construction of piles. Supervision should be provided by both the piling contractor and the Engineer and also where applicable by the Contractor.

## 3.6. Non-conformance

Sometimes events occur which cannot be avoided or are not confirmed until one or more elements have been constructed. In these circumstances the procedure below is required to consider the acceptability of the work. (For example, see Clause 3.2.)

If any part or parts of the foundation are discovered not to conform with the standard of materials, workmanship or design (where appropriate) required under the Specification then

(*a*) The Contractor shall, within seven days of discovery or receipt of the Engineer's request for further information, notify the Engineer of his proposals and provide calculation or additional information necessary to remedy the non-conformance or substantiate that the work done is of equivalent value and effectiveness.

(*b*) The Engineer shall be deemed to have accepted that the proposal demonstrates its ability to satisfy the intent of the Specification unless within seven days of receipt of such proposals he details a request for any additional information for this purpose.

Where a non-conformance is allowed and confirmed

(*a*)   an appropriate deduction may be made in the adjustment of the Contract Sum

and/or

(*b*)   such Variations shall be issued as are reasonably necessary as a consequence of the above but no addition to the Contract Sum shall be made and no extension of the time shall be given.

The above clause is suggested to encourage the rapid resolution of difficulties on site and to clarify the position of the parties to the contract.

# 4. Use of the *Specification for piling and embedded retaining walls*

## 4.1. Application

The *Specification for piling and embedded retaining walls* applies to bearing piles, and is in accordance with the general principles of BS 8004 — *Foundations*. The *Specification for piling and embedded retaining walls* covers materials and workmanship and does not stipulate design criteria.

The Engineer may use the specification in two ways

(*a*) by referring in the contract documents to the whole, or complete sections of it, and including in the contract documents a statement of amendments and a Particular Specification

(*b*) by including in the contract documents the whole, or complete sections of it, together with a statement of amendments and a Particular Specification.

Every effort has been made to avoid conflict between the specification, the ICE conditions, the JCT conditions and other standard forms of contract. However, certain clauses in the ICE conditions do not have parallel clauses in other conditions. Therefore, clauses in the specification which the Engineer considers are covered by the conditions or the specification for the (main) contract (if appropriate) must be amended or deleted.

Where a given type of pile is to be specified the appropriate sections only need to be used (e.g. for driven cast-in-place piles, sections 1, 5, 8, 9, 10 and 20). However, in many instances it can be advantageous to have the benefit of the specification as a whole, so as to provide for any unexpected developments which might lead to a change in the type of pile.

## 4.2. Particular Specification clauses

All sections except Section 20 of the *Specification for piling and embedded retaining walls* require clauses to be given in a Particular Specification to cover specific requirements for a particular type of pile, wall element or testing as noted in the relevant clauses. It is desirable that the clauses of a Particular Specification should be drafted and set out in conformity with the format of the *Specification for piling and embedded retaining walls*.

## 4.3. Concrete in piles

Although precast piles conform to the requirements of codes of practice and British Standards relating to concrete mix design, difficulty may arise in meeting such requirements in the case of cast-in-place piles due to the special requirements necessary to ensure the integrity of the pile shaft. It is therefore to be preferred in the case of cast-in-place piling, notwithstanding the requirements of the relevant British Standards, that concrete mixes should meet the requirements of the *Specification for piling and embedded retaining walls* where there is conflict with British Standards.

In the case of bored and cast-in-place driven piles, particular attention needs to be given to special requirements such as high workability and coherence in order to ensure the integrity of a pile

*Specification for piling and embedded retaining walls*. Thomas Telford, London, 1996.

129

shaft. Such criteria may govern acceptable levels of cement content etc., and may pre-empt other requirements of concrete codes. High characteristic concrete strength requirements can exacerbate thermal (shrinkage) cracking of the pile shaft, which may be revealed by integrity-testing. It may be necessary to introduce sufficient reinforcement to resist these contraction stresses, although the alternative of introducing low heat cements should be considered since this will normally offer a more advantageous solution provided that due consideration is given to associated reductions in gain of strength at 28 days.

Very high strength concrete in pile shafts can also lead to difficulties in achieving satisfactory trimming of pile heads without the risk of secondary damage to the shafts of piles below cut-off level.

The treatment of concrete strength acceptability on the basis of a statistical approach is quite normal for factory-produced precast concrete piles. However, on many cast-in-place piling sites, the volume of concrete used is relatively small and the duration of the works is short and in such cases the statistical approach cannot be justified.

## 4.4. Tolerances

Clauses 1.8 and 11.8 of the *Specification for piling and embedded retaining walls* stipulate tolerances which are realistic for most sites and ground conditions. It should be noted that where the ground contains obstructions, tolerances may need to be increased and appropriate allowances made in the initial design of the pile caps to accommodate such increased tolerances. Ground beams may need to be redesigned to suit the positions of piles as finally installed. The need for close tolerances diminishes with capped groups of three or more piles. Specific requirements for difficult sites should be stated in the Particular Specification.

## 4.5. Pile testing procedures

It is important to differentiate between load-testing and integrity-testing. Guidance on the two forms of testing is provided in Appendices D, E and F.

Load-testing entails the measurement of the response of the head of the pile but provides no assurance that the construction of the pile is sound nor that it meets the requirements of a specification. The elastic response of the pile-soil or pile-rock system should be recognized and no criteria for movement at the head of the pile should be specified which are less than the elastic response of the system to applied load. The possibility of inelastic movements occurring, as the ratios of applied stress to limiting stress of the soil or rock increase, must also be recognized in the specification of criteria for acceptance of pile performance.

Dynamic testing entails the measurement of the immediate dynamic response of the pile-soil or pile-rock system to the application of a dynamic load, commonly provided to the head of a pile by a piling hammer. An analysis of the response will normally be carried out after the test.

Integrity-testing entails the measurement of a property of a pile which can be related to its soundness but does not provide any assurance that the pile is capable of safely supporting the specified working load.

In the event that integrity tests indicate potential defects in a pile and that other subsequent tests prove the piles to be defective, then the costs of all further testing, investigation, remedial works and/or replacement of the defective pile should be borne by the Contractor. If, following investigation, the subsequent tests do not reveal significant defects, the cost of such further tests, investigations and reinstatement should be borne by the Employer.

## 4.6. Degaussing steel piles

Magnetization of the heads of driven piles can occur and may result in magnetic arc blow (i.e. displacement of the welding arc) during butt-welding of pile extension pieces, with detrimental effects on the quality of weld.

Degaussing of the pile head is necessary in these circumstances and can be achieved by the generation of a counteractive magnetic field during the period when welding is in process.

## 4.7. Continuous flight auger piles

The method of construction of continuous flight auger piles requires the reinforcement cage to be inserted into the pile after completion of the concreting/grouting operation. There are therefore practical limitations to the length of reinforcement cage which can be inserted and it may not be practicable to reinforce long piles for their full length.

# 5. Measurement of piling and embedded retaining wall work

## 5.1. General

Bills of quantities for building and civil engineering works are normally prepared in conformity with a standard method of measurement approved for use with the relevant form of contract. Some of the provisions of these standard methods are not, however, considered to be well suited to the efficient and practical measurement of piling. Outline bills of quantities appropriate to different types of piling are included in Appendix A. The principles set out therein are commended to future revision committees of standard methods of measurement and to those responsible for measuring piling works.

The following paragraphs explain the intention of the suggested items included in the outline bills of quantities given in Appendix A. There will need to be variations or additions to these items to meet particular circumstances, structures or sites.

Materials from which the piles are formed and the section characteristics of these piles should be stated in item descriptions. Section characteristics are

(*a*) for steel piles, section reference or mass per metre and cross-sectional dimensions
(*b*) for other than steel piles, cross-sectional dimensions or specified diameters.

It should be noted that the finished diameter of a properly constructed cast-in-place pile may be greater than but not less than the specified diameter due to the method of construction or the nature of the ground.

Piling work should always be billed as 'provisional' and measured and valued as executed since it is only in exceptional circumstances that piles will be installed to the precise depths shown on the drawings or as measured. Provision for contingencies should be made by providing provisional sums in the bill of quantities and not by increasing the quantities beyond those expected to relate to the work to be executed.

The quantities should be computed net from the drawings unless directed otherwise in the Contract. No allowance should be made in the measurement of a pile for an increase in the specified diameter, waste or cut-off level tolerances.

## 5.2. Preliminaries

The model provisions for incorporation into piling tenders, and the typical schedule of attendances and facilities which would commonly be provided by a Contractor to a piling contractor are set out in Appendices B and C and are not therefore shown in the list of measured items (Appendix A).

Separate measurement items should be included for each required establishment on site, distinguishing between that for any preliminary piles and for the main piling works. Unless there is a particular requirement, the establishment of plant on site and its subsequent removal can be measured as one item.

Additional items for establishment within the site should be included when piling is to be installed in separate work areas which require the transfer of units of plant involving an expenditure of time and resources significantly greater than that required for moving from pile position to pile position within the same area. The separate areas should be defined on the drawings and in the Particular Specification or the bill of quantities.

The terms and conditions of the standard forms of contract set down the obligations of the Employer and the Contractor to each other in respect of the provision of insurances. Where the Contractor has to effect insurance cover, the Employer's requirements must be set out in the contract documents. Should there be no direct obligations included in a contract, such as GC/Works/1 form of contract, for the Contractor to effect insurances to meet contractual liabilities, he is still obliged to obtain insurance cover to comply with statutory requirements, and it is advisable that he should take out appropriate insurance cover as a matter of commercial prudence.

In all standard forms of contract the responsibility for the accurate setting-out of the structure rests with the Contractor. If it is specified that the setting-out of the pile positions is the responsibility of a piling sub-contractor, the Contractor must provide and maintain permanent stations setting out the positions of the structural grid lines and temporary benchmarks, and be responsible for maintaining them until he has satisfied himself that the piles have been installed in the correct positions, or, if a pile is found to be incorrectly positioned, until the piling sub-contractor has been notified and given the opportunity to verify the alleged error.

Item descriptions in the bills of quantities should identify work which is affected by bodies of water (other than groundwater) such as rivers, streams, canals, lakes and tidal water. Item descriptions for work affected by tidal water should also distinguish between work affected at all times and work affected only at some states of the tide. Water surface levels adopted for the purpose of such distinctions should be stated in item descriptions.

### 5.3. Obstructions

The removal of overhead, surface and underground obstructions and the backfilling of the voids with suitable material, through which piles can be readily driven or bored and soundly constructed, will need to be carried out by the Contractor before piling work begins (see Appendix C).

### 5.4. Working levels

The piling platform level is the level at which the piling rig stands to carry out the work. Where necessary the piling platform and access thereto will be suitably prepared to support adequately the piling plant and equipment and attendant transport.

Where piling has to be carried out from a fixed structure or staging, whether permanent or temporary, or where floating or temporary staging may be used at the Contractor's discretion, the bill of quantities should be drafted accordingly.

'Commencing surface' in relation to an item in a bill of quantities is the surface level at the pile position at the commencement of the boring or driving operation. Work from each commencing surface which differs from the piling platform level and where the Contractor may incur significant additional costs should be billed separately (see Fig. 1).

On land the commencing surface will normally be the piling platform level; it can, however, be higher than the piling platform level when the pile is formed through a berm or the slope of an

*Specification for piling and embedded retaining walls.* Thomas Telford, London, 1996.

133

embankment, or lower when working adjacent to an excavation or cofferdam (see Fig. 2). In cases where the rig working level is above the commencing surface (e.g. in marine works from a staging or in a cofferdam with the rig at the top of the excavation), the bored or driven pile length which will be paid for should be measured from the bed level or the bottom of the excavation (see paragraphs 5.6.1, 5.9.1 and 5.10.1 below).

## 5.5. Measurement of lengths and depths

The concreted length of a cast-in-place pile is the length of the in-situ concrete measured from the specified cut-off level or the commencing surface, whichever is lower, to the specified toe level of the pile. Where pile cut-off levels are higher than the commencing surface, it may be necessary to extend the piles. The unconcreted length (empty bore/drive) between the specified cut-off and the commencing surface should be included in the item for total depth bored or driven. In this connection attention is directed to the provisions of the *Specification for piling and embedded retaining walls* for casting level tolerances in respect of cast-in-place piles.

The length of preformed concrete, timber or steel piles is the length stipulated or approved by the Engineer to be supplied, handled and pitched prior to driving. This length does not include unforeseen extensions. When driving to a set which is not achieved within the expected length, the Contractor should have the Engineer's written instruction before any pile is extended. The nature of preformed piles demands that the Contractor is instructed as to the length to be supplied to the pile position in sufficient time to enable the length to be cast or purchased.

Bored and driven depths should be measured from the commencing surface to the bottom of the casings of driving cast-in-place piles and to the toe level of other piles, with due adjustment for raking piles.

Separate items, similar to those for vertical piles, should be provided in a bill of quantities for raking piles, the amount of the rake being stated in the description. Measurement should be along the axes of the piles. In practice there is a limit to the amount of rake to which some types of pile can be installed.

## 5.6. Cast-in-place bored and driven piles
### 5.6.1. General

Driven permanently cased piles formed by driving a steel casing or concrete shell in one or more pieces which remain in place after driving and which are filled with concrete, are included in this section.

The following separate items are required in a bill of quantities for different work areas and commencing surfaces and for each pile diameter detailed on the drawings

(*a*)  the number of piles
(*b*)  the total depth of the bored or driven piles (the maximum anticipated length of the piles is to be stated)
(*c*)  the total concreted length of piles.

The maximum depth to be bored or driven will influence a contractor in his selection of the plant and equipment to be used and this in turn will affect the price offered. Plant and equipment do possess limitations in performance, so variation orders which increase the tendered maximum depth may produce a situation where the capacity of the plant and equipment upon which the contract is based may have to be increased, designs changed, or other piling systems substituted (see Fig. 3). To overstate the depth may produce an uneconomic tender due to the capacity of the plant offered being in excess of that which is required.

The forming of an enlarged base to a driven cast-in-place pile by driving out a concrete plug should be measured as an extra-over item. The diameter of the shaft should be given in the item description, but not the size of the enlarged base as this cannot be predetermined and is not subject to remeasurement.

### 5.6.2. Casings

Where permanent casings are specified as part of the piling works, these should be measured. The internal diameter and performance criteria for the casing should be contained in the item description. If anti-friction coating or other applied finishes are required on the outside surface, this requirement should be included in the item description.

Sufficient time must be allowed for permanent casings to be manufactured and delivered to site in time for the commencement of the piling work. As this material forms part of the permanent works, it is incumbent upon the Engineer to specify the length which has to be placed at a pile position.

Where temporary casings are left in as a result of unforeseen ground conditions and on the specific instructions of the Engineer, the additional costs incurred by the Contractor should be reimbursed.

### 5.6.3. Cutting-down of pile heads

The extent of cutting-down of the heads of cast-in-place piles will be governed by the depth of the specified cut-off level below the commencing surface, the standing water level if any, the length of the temporary casings required to support the sides of the bore, and whether or not concrete has been placed under fluid. An item for cutting down the heads of cast-in-place piles, disposal of the debris and preparation to receive the pile caps should be measured taking into account the cut-off level tolerances stated in the specification. It is normally economical for the Contractor to trim the piles and dispose of surplus lengths of piles.

### 5.6.4. Material for disposal

The volume of surplus excavated material for disposal should be calculated net from the specified cross-sectional area of the bored cast-in-place piles, or of the preboring for driven cast-in-place piles, and the bored lengths as measured in accordance with the Contract. The volume of enlarged bases to bored cast-in-place piles should be added. The actual volume as excavated can be greater due to the use of temporary casings and the nature of the ground.

A provisional item for excavating or re-excavating oversite to reduced level and disposal of the surplus material should be included in the appropriate bill of quantities to allow for possible additional excavation required due to any upward displacement of ground brought about by the driving of cast-in-place or preformed piles.

### 5.7. Special requirements for cast-in-place bored piles

The forming of an enlarged base to a bored cast-in-place pile by underreaming is measured as extra over to the shaft. The diameter of the shaft and the diameter of the enlarged base should be given in the item description. An item should be included in the bill of quantities to cover the provision of the additional equipment and attendance necessary for the inspection of the pile bores and underreams.

If a pile is to be founded on or in rock, the nature and properties of the founding stratum must be defined in the description and, in the case of bored cast-in-place piles, the amount of penetration should be stated and included in the depth of boring.

As boulders and discontinuous rock inclusions occurring in a shaft cannot be measured with any degree of precision, the payment for penetrating such strata should be treated as an obstruction and paid for on an hourly basis. The type of plant and the method to be utilized for the removal of the obstruction should be stated.

Where concrete is specified to be placed under water or a bentonite suspension as the prescribed method of construction, reference must be made in the bill to this additional work. However, it need not be a separate item but can be included in the item description for the concreted length of pile. The length measured for this work should be the full length of the concreted pile, even if the bore is only partly filled with fluid.

An item for placing concrete by means of a tremie should not be included in a bill of quantities as a contingency item or just to obtain a rate for the work, as additional plant, equipment and labour would be required, the concrete would need to be enriched and the rate of production would be reduced.

When bored cast-in-place piles are installed through water or a bentonite suspension, the Contractor needs to take this into account when disposing of the arisings from the bore, as they will be wet, or contaminated by the bentonite suspension.

Provision will be required on site for the disposal of water arising from the bores as it is displaced by the concrete. The disposal of bentonite suspension should be described in the item for boring.

## 5.8. Measurement of reinforcement in cast-in-place piles

A bill of quantities should describe the kind and quality of the steel and the section of the bars if other than plain circular in cross-section. Each bar diameter should be given separately and a distinction should be made between straight bars, lateral ties and helical reinforcement. The rate for reinforcement should include steel in laps, tying wire and additional steel for handling purposes to the extent required for the work detailed in the tender drawings. It should also include spacers, which need to be specified as to size, quality and frequency unless these requirements are left to the discretion of the (piling) contractor. The projecting reinforcement required for building into the pile cap should be included in the measurement.

Reinforcement in raking piles should be billed separately from that in vertical piles.

An alternative to measuring reinforcement separately which may be adopted for piles of diameter up to 600 mm is to describe the reinforcement required in the item for the concreted length of pile. This method is appropriate to small-diameter piles which are specified to be reinforced for their full length since the cost of the reinforcement is not a significant item when compared against the total cost of the pile.

## 5.9. Precast concrete, timber and steel piles
### 5.9.1. General

The following separate items are required in a bill of quantities for different work areas and commencing surfaces and for each cross-section or mass per metre detailed on the drawings

(a)  the number of piles
(b)  the length of each pile and the number to be supplied

(*c*)   the total depth driven (the maximum anticipated depth of the piles should be stated).

If slip coating or other applied finishes are required to be applied to the surface to reduce downdrag or to protect the pile, this requirement should be included as a separate item description for the length of piles required.

### 5.9.2. Length of pile supplied

The actual supply length of a preformed pile will affect the unit rate, as the most economic casting lengths of precast concrete piles will incorporate standard lengths of steel as offered by the manufacturers.

Where there is a requirement for long single-length precast concrete piles, normally in excess of 13 m long, prestressed concrete piles will often be specified.

Steel piling is supplied by manufacturers in lengths under 3 m, 3 m to under 5 m, 5 m to 14 m, over 14 m to 24 m, over 24 m to 30 m, and over 30 m, the purchase price per tonne being governed by the purchased length.

### 5.9.3. Pile extensions

Pile extensions should not be included in the measurement of the main piling but billed as separate items of work. There are three items required to describe the work to an extension: a numbered item, a pile unit length item and a depth driven item. Contractors do not have a uniform approach to the pricing of their fixed costs for extensions, as the nature of these costs permits them to be allocated either against the numbered item or within the billed unit length. These two items should not be subject to variation once instructed, as is the case with the driving item. The numbered item should include for preparing the head to receive the extension, handling and pitching and making good applied coatings or finishes; additionally, in the case of a steel pile, the welding-on of the extension.

Steel piles can be extended by welding on a length that has been cut from a pile which has not been driven to its full supply length. Degaussing is seldom necessary and should not be regarded as included but should be measured as a provisional extra-over item. Cut-off lengths from precast concrete piles cannot be reused.

An alternative method of measurement for the numbered item and supply length is to bill extensions as a numbered item stating the length required to be supplied and including all the labour and other items associated with an extension.

### 5.9.4. Applied finishes

If an anti-friction coating or other applied finish is required on the outside surface, this requirement should be included in the item description.

### 5.9.5. Cutting-down of pile heads

The extent of the cutting-down of the heads of preformed piles will be the length of the pile from the specified cut-off level to the head of the supply length; this item should include for exposing the steel reinforcement at the pile cut-off level and disposal of the debris. In respect of steel piles, the ownership of the cut-off lengths and responsibility for their disposal should be stated. Should the Contractor supply a preformed pile of a greater length than that specified or ordered by the Engineer, this extra length remains the property of the Contractor.

### 5.9.6. Rock shoes

Where precast concrete and steel piles are required to be fitted with rock shoes, the type and weight should be specified.

### 5.9.7. Preboring

Where it is a requirement to prebore for a preformed pile, the Engineer should specify the method, and if this results in a void between the sides of the excavation and the pile, an item should be provided for infilling with approved material.

### 5.10. Jointed precast concrete segmental piles
### 5.10.1. General

The following separate items are required in a bill of quantities for different work areas and commencing surfaces and each cross-section detailed on the drawings

(a)  the number of piles
(b)  the length of each pile toe unit and the number to be supplied
(c)  the length of each extension unit and the number to be supplied
(d)  the number of pairs of mechanical joints to be supplied
(e)  the total depth to be driven.

If slip coating or other applied finishes are required to be applied to the surface to reduce downdrag or to protect the pile, this requirement should be included as a separate item description for the length of piles required.

### 5.10.2. Length of pile

The maximum length of the pile which can be transported by road as a normal load is 13 m. In the case of precast jointed concrete piles, lengths in excess of 13 m are normally catered for by the addition of mechanically jointed extension units.

### 5.10.3. Pile extensions

Should the Engineer require provision for an extension to a length of jointed precast concrete segmental pile, instructions prior to casting must be given for a mechanical joint to be fitted to the head of the pile toe unit. To cater for this possibility, it is prudent to include an item in the bill of quantities.

Where there is a requirement to extend a segmental pile which has been supplied without a mechanical joint, a solution should be adopted to suit the loading of the pile. Examples may include pile-trimming to expose reinforcement, with in-situ extension, and epoxy resin or similar jointing to attach a further precast pile segment. An item for this work should be included in the bill of quantities.

### 5.11. Preliminary piles

Preliminary piles installed and tested before the start of the main works should be measured in a separate bill of quantities or be the subject of a separate contract. Their installation should be measured in accordance with the appropriate method of measurement for the type of pile, and separate items should be measured for each specified loading and type of test. The specification should state whether preliminary piles are to remain as constructed, or the tops are to be cut off, or if they are to be incorporated into the Works or to be withdrawn.

### 5.12. Tests on working piles

Separate items should be provided in the bill of quantities for each specified loading. The item description should include all preparatory work such as site surfacing, bringing the pile head to the commencing surface level, trimming the head of the test pile and all work required in connection with the form of anchorage used including, where applicable, the supply of kentledge.

### 5.13. Plant standing during tests

If plant is required to stand between tests on preliminary piles and the start of the main piling contract work, during tests on working piles or while soil samples are taken, payment should be subject to the rules applying to daywork or on the basis of an itemized delay rate.

### 5.14. Samples and in-situ tests

Separate items should be provided in the bill of quantities for each form of sampling and testing operations; for example, sampling and testing undisturbed soil samples, penetration tests, manufacture and testing of concrete cubes, integrity-testing and dynamic load testing.

### 5.15. Removal of artificial and natural obstructions

The removal of natural obstructions such as boulders and discontinuous rock inclusions and unquantifiable artificial obstructions occurring within the shaft of a pile should be allowed for by provisional items in the bill of quantities, the Contractor being reimbursed for overcoming such obstructions.

Payments should be subject to the rules applying to daywork and should be paid extra over the measured work for the piles involved.

### 5.16. Payment on daywork basis

Provision should be made in an appendix to the bill of quantities for the Contractor to define the equipment on which his tender is based, together with the hourly rates for each plant and gang.

### 5.17. Observations and records

The cost of preparation and submission of records prescribed in clauses 1.11 and 11.11 and Tables 1.1 and 11.1 of the *Specification for piling and embedded retaining walls* should be included in the rates. These records are important, as this information will be used for the purpose of measurement of the Works and the preparation of as-made records of the Works.

### 5.18. Outline bills of quantities

The following outline bills for different types of piling are intended to illustrate the recommended items of measurement in accordance with the foregoing recommendations but are not intended to be comprehensive and do not, for example, include raking piles and provisional items.

*Specification for piling and embedded retaining walls.* Thomas Telford, London, 1996.

139

# Appendix A.
## Model bills of quantities

### Bored cast-in-place piles

| Number | Item description | Unit | Quantity | Rate | Amount £ p |
|---|---|---|---|---|---|
| A | Transport piling plant and equipment to site, set up, dismantle and remove upon completion.   Per visit | sum | | | |
| B | Move between area * and area * | sum | | | |
| | *The following work to be carried out from a piling platform level of * m OD with a Commencing Surface level of * m OD* | | | | |
| C | Movement of plant and equipment to each pile position, including setting up rig<br>    * mm specified diameter<br>    * mm specified diameter | <br><br>nr<br>nr | | | |
| D | Bore pile shafts to depths not exceeding * m<br>    * mm specified diameter<br>    * mm specified diameter | <br>m<br>m | | | |
| E | Concrete quality * in pile shafts<br>    * mm specified diameter<br>    * mm specified diameter | <br>m<br>m | | | |
| F | Enlarged bases including forming and concrete quality * therein<br>    Shaft dia.   Base dia.<br>    * mm        * mm<br>    * mm        * mm | <br><br><br>nr<br>nr | | | |
| G | Plain grade 250 straight bars in reinforcing cage<br>    * mm diameter<br>    * mm diameter | <br>t<br>t | | | |
| H | Plain grade 250 helical binding * mm diameter at * mm pitch | t | | | |
| J | Plain grade 250 permanent casing * mm wall thickness<br>    * mm internal diameter (in * nr)<br>    * mm internal diameter (in * nr) | <br><br>m<br>m | | | |
| K | Backfill open shafts with *, as specified in the particular specification | m³ | | | |
| L | Remove excavated material from the area around the pile position during boring operations, including loading and depositing * | m³ | | | |
| M | Cut down concrete pile shaft to the specified cut-off level, prepare exposed head and reinforcement to receive capping: load and dispose of debris<br>    * mm specified diameter<br>    * mm specified diameter | <br><br><br>nr<br>nr | | | |
| N | Make concrete cube and test | nr | | | |
| P | Take undisturbed sample 100 mm diameter from bore at depths not exceeding * m | nr | | | |
| Q | Laboratory tests: set of three undrained triaxial compression tests on 100 mm samples, including handling and transport | nr | | | |
| R | Standard penetration test, depth not exceeding * m | nr | | | |
| S | Provide equipment and labour to descend shafts for inspection purposes | sum | | | |
| T | Transport equipment to site for maintained load tests not exceeding * kN and remove upon completion of tests | sum | | | |

*Specification for piling and embedded retaining walls*. Thomas Telford, London, 1996.

| Number | Item description | Unit | Quantity | Rate | Amount £ p |
|--------|-----------------|------|----------|------|------------|
| V | Maintained load test on a * mm diameter working pile to * kN, including reaction | nr | | | |
| W | Cut down * mm pile cap and tops of reaction piles to a depth of * m , dispose of debris, backfill, and make good disturbed site surface | nr | | | |
| X | Carry out approved integrity tests (minimum 20 per visit) and provide report by approved specialist | nr | | | |
| Y | Remove obstruction in bore, utilizing standard equipment on site. Each plant and gang | h | | | |
| Z | Charge for plant and labour delayed by causes beyond the piling contractor's control or when instructed. Each plant and gang | h | | | |

## Driven cast-in-place piles

| Number | Item description | Unit | Quantity | Rate | Amount £ p |
|--------|-----------------|------|----------|------|------------|
| A | Transport piling plant and equipment to site, set up, dismantle and remove upon completion.    Per visit | sum | | | |
| B | Move between area * and area * | sum | | | |
| | *The following work to be carried out from a piling platform level of * m OD with a Commencing Surface level of * m OD* | | | | |
| C | Movement of plant and equipment to each pile position, including setting up rig and handling and pitching casing | nr | | | |
| D | Drive pile casing to depths not exceeding * m | m | | | |
| E | Concrete quality * in pile shafts * mm specified diameter | m | | | |
| F | Enlarged base * mm diameter to shafts of * mm specified diameter | nr | | | |
| G | Plain grade 250 straight bars in reinforcing cage; * mm diameter | t | | | |
| H | Plain grade 250 helical binding * mm diameter at * mm pitch | t | | | |
| J | Mild steel permanent casing * mm wall thickness and * mm internal diameter (in * nr) | m | | | |
| K | Transport preboring machine to site and remove upon completion | sum | | | |
| L | Prebore for * mm specified diameter piles and deposit excavated material adjacent to head of shafts | m | | | |
| M | Backfill open shafts with *, as specified in the particular specification | m³ | | | |
| N | Remove excavated material from the area around the pile position during boring operations, including loading and depositing * | m³ | | | |
| P | Cut down * mm specified diameter concrete pile shaft to the specified cut-off level, prepare exposed head and reinforcement to receive capping; load and dispose of debris | nr | | | |
| Q | Make concrete cube and test | nr | | | |
| R | Transport equipment to site for maintained load tests not exceeding * kN and remove upon completion of tests | nr | | | |
| S | Maintained load test on a * mm diameter working pile to * kN, including reaction | nr | | | |
| T | Carry out approved integrity tests (minimum 20 per visit) and provide report by approved specialist | nr | | | |
| V | Drive through obstructions, utilizing standard equipment on site. Each plant and gang | h | | | |
| W | Charge for plant and labour delayed by causes beyond the piling contractor's control or when instructed. Each plant and gang | h | | | |

# Jointed precast concrete segmental piles

| Number | Item description | Unit | Quantity | Rate | Amount £    p |
|---|---|---|---|---|---|
| A | Transport piling plant and equipment to site, set up, dismantle and remove upon completion.  Per visit | sum | | | |
| B | Move between area * and area * | sum | | | |
| | *The following work to be carried out from a piling platform level of * m OD with a Commencing Surface level of * m OD* | | | | |
| C | Movement of plant and equipment to each pile position, including setting up rig and handling and pitching pile segments | nr | | | |
| D | Supply * m long precast concrete segmental pile toe units * mm × * mm in cross-section | m | | | |
| E | Supply * m long extension units * mm × * mm in cross-section | m | | | |
| F | Supply cast -in mechanical joints (pairs) | nr | | | |
| G | Prepare pile heads for extension using approved methods and materials | nr | | | |
| H | Pitch and drive piles to depths not exceeding * m | m | | | |
| J | Credit for unused half joints | nr | | | |
| K | Transport preboring machine to site and remove upon completion | sum | | | |
| L | Prebore for pile with * mm × * mm cross-section and deposit excavated material adjacent to head of shaft | m | | | |
| M | Remove excavated material from the area around the pile position during preboring operations, including loading and depositing at *.......................... | m³ | | | |
| N | After driving pile, fill annulus of prebore with *............................. | m | | | |
| P | Cut-off surplus pile * mm × * mm in cross-section, dispose of debris and prepare head and reinforcement to receive capping | nr | | | |
| Q | Transport equipment to site for maintained load tests not exceeding * kN and remove upon completion of tests | sum | | | |
| R | Maintained load test on a working pile to * kN, including reaction | nr | | | |
| S | Driving through obstructions, utilizing standard equipment on site. Each plant and gang | h | | | |
| T | Charge for plant and labour delayed by causes beyond the piling contractor's control or when instructed. Each plant and gang | h | | | |

*Specification for piling and embedded retaining walls.* Thomas Telford, London, 1996.

## Driven steel piles

| Number | Item description | Unit | Quantity | Rate | Amount £ p |
|---|---|---|---|---|---|
| A | Transport piling plant and equipment to site, set up, dismantle and remove on completion.   Per visit | sum | | | |
| B | Move between area * and area * | sum | | | |
| | *The following work to be carried out from a piling platform level of * m OD with a Commencing Surface of * m OD* | | | | |
| C | Movement of plant and equipment to each pile position, including setting up rig and handling and pitching pile | nr | | | |
| D | Supply * mm × * mm steel * section piles (* kg/m), each * m long | m | | | |
| E | Supply * mm × * mm extension piles (* kg/m), each * m long | m | | | |
| F | Provide covered structure under which coating of piles is undertaken, and remove on completion | sum | | | |
| G | Blast-cleaning * mm × * mm pile | m² | | | |
| H | Paint * coats of primer and * coats of * paint, the whole having a minimum total thickness of * mm | m² | | | |
| J | Pitch and drive piles to depths not exceeding * m | m | | | |
| K | Provide site-welded butt joints between piles and extensions; * mm × * mm | nr | | | |
| L | Cut off surplus lengths after driving * mm × * mm piles (* kg/m) and stack for reuse where authorized | nr | | | |
| M | Transport preboring machine to site and remove upon completion | sum | | | |
| N | Prebore for steel pile with * mm × * mm cross-section and deposit excavated material adjacent to head of shaft | m | | | |
| P | Remove excavated material from the area around the pile position during preboring operations, including loading and depositing at *........................ | m³ | | | |
| Q | After driving pile, fill annulus of prebore with *............................. | m | | | |
| R | Cut-off surplus steel pile * mm × * mm in cross-section, dispose of debris and prepare head to receive capping | nr | | | |
| S | Drive through obstructions, utilizing standard equipment on site. Each plant and gang | h | | | |
| T | Charge for plant and labour delayed by causes beyond the piling contractor's control or when instructed. Each plant and gang | h | | | |

## Contiguous bored pile walls

| Number | Item description | Unit | Quantity | Rate | Amount £ p |
|---|---|---|---|---|---|
| A | Transport piling plant and equipment to site, set up, dismantle and remove upon completion.   Per visit | sum | | | |
| B | Move between area * and area * | sum | | | |
| | *The following work to be carried out from a piling platform level of * m OD with a Commencing Surface level of * m OD* | | | | |
| C | Movement of plant and equipment to each pile position, including setting up rig | | | | |
| | * mm specified diameter | nr | | | |
| | * mm specified diameter | nr | | | |
| D | Bore pile shafts to depths not exceeding * m | | | | |
| | * mm specified diameter | m | | | |
| | * mm specified diameter | m | | | |

| Number | Item description | Unit | Quantity | Rate | Amount £ p |
|--------|-----------------|------|----------|------|------------|
| E | Concrete quality * in pile shafts<br>  * mm specified diameter<br>  * mm specified diameter | m<br>m | | | |
| F | Deformed grade 460 straight bars in reinforcing cage<br>  * mm diameter<br>  * mm diameter | t<br>t | | | |
| G | Plain grade 250 helical binding * mm diameter at * mm pitch | t | | | |
| H | Plain grade 250 permanent casing * mm wall thickness<br>  * mm internal diameter (in * nr)<br>  * mm internal diameter (in * nr) | m<br>m | | | |
| J | Backfill open shafts with *, as specified in the particular specification | m³ | | | |
| K | Remove excavated material from the area around the pile position during boring operations, including loading and depositing * | m³ | | | |
| L | Cut down concrete pile shaft to the specified cut-off level, prepare exposed head and reinforcement to receive capping: load and dispose of debris<br>  * mm specified diameter<br>  * mm specified diameter | nr<br>nr | | | |
| M | Make concrete cube and test | nr | | | |
| N | Take undisturbed sample 100 mm diameter from bore at depths not exceeding * m | nr | | | |
| P | Laboratory tests: set of three undrained triaxial compression tests on 100 mm samples, including handling and transport | | | | |
| Q | Standard penetration test, depth not exceeding * m | nr | | | |
| R | Carry out approved integrity tests (minimum 20 per visit) and provide report by approved specialist | nr | | | |
| S | Remove obstruction in bore, utilizing standard equipment on site. Each plant and gang | h | | | |
| T | Charge for plant and labour delayed by causes beyond the piling contractor's control or when instructed. Each plant and gang | h | | | |

# Secant bored pile walls

| Number | Item description | Unit | Quantity | Rate | Amount £ p |
|--------|-----------------|------|----------|------|------------|
| A | Transport piling plant and equipment to site, set up, dismantle and remove upon completion.    Per visit | sum | | | |
| B | Move between area * and area *<br>*The following work to be carried out from a piling platform level of * m OD with a Commencing Surface level of * m OD* | sum | | | |
| C | Movement of plant and equipment to each pile position, including setting up rig<br>  * mm specified diameter<br>  * mm specified diameter | nr<br>nr | | | |
| D | Bore female pile shafts to depths not exceeding * m<br>  * mm specified diameter<br>  * mm specified diameter | m<br>m | | | |
| E | Bore male pile shafts to depths not exceeding * m<br>  * mm specified diameter<br>  * mm specified diameter | m<br>m | | | |
| F | Concrete or bentonite/cement mix quality * in female pile shafts<br>  * mm specified diameter<br>  * mm specified diameter | m<br>m | | | |
| G | Concrete quality * in male pile shafts<br>  * mm specified diameter<br>  * mm specified diameter | m<br>m | | | |

*Specification for piling and embedded retaining walls.* Thomas Telford, London, 1996.

| Number | Item description | Unit | Quantity | Rate | Amount £ p |
|--------|------------------|------|----------|------|------------|
| H | Deformed grade 460 straight bars in reinforcing cage<br>    * mm diameter<br>    * mm diameter | t<br>t | | | |
| J | Plain grade 250 helical binding * mm diameter at<br>* mm pitch | t | | | |
| K | Plain grade 250 permanent casing * mm wall<br>thickness<br>    * mm internal diameter (in * nr)<br>    * mm internal diameter (in * nr) | <br><br>m<br>m | | | |
| L | Structural steel section * in pile shaft | m | | | |
| M | Backfill open shafts with *, as specified in the<br>particular specification | m³ | | | |
| N | Remove excavated material from the area around the<br>pile position during boring operations, including<br>loading and depositing * | m³ | | | |
| P | Cut down concrete pile shaft to the specified cut-off<br>level, prepare exposed head and reinforcement to<br>receive capping: load and dispose of debris<br>    * mm specified diameter<br>    * mm specified diameter | <br><br><br>nr<br>nr | | | |
| Q | Make concrete cube and test | nr | | | |
| R | Take undisturbed sample 100 mm diameter from<br>bore at depths not exceeding * m | nr | | | |
| S | Laboratory tests: set of three undrained triaxial<br>compression tests on 100 mm samples, including<br>handling and transport | nr | | | |
| T | Standard penetration test, depth not exceeding * m | nr | | | |
| V | Carry out approved integrity tests (minimum 20 per<br>visit) and provide report by approved specialist | nr | | | |
| W | Remove obstruction in bore, utilizing standard<br>equipment on site. Each plant and gang | h | | | |
| X | Charge for plant and labour delayed by causes<br>beyond the piling contractor's control or when<br>instructed. Each plant and gang | h | | | |

## Diaphragm walls

| Number | Item description | Unit | Quantity | Rate | Amount £ p |
|--------|------------------|------|----------|------|------------|
| A | Transport diaphragm walling plant and equipment to<br>site, set up, dismantle and remove upon completion.<br>Per visit | sum | | | |
| B | Move between area * and area * | sum | | | |
| | *The following work to be carried out from a piling<br>platform level of * m OD with a Commencing<br>Surface level of * m OD* | | | | |
| C | Movement of plant and equipment to each panel<br>position, including setting up rig<br>    * mm specified thickness<br>    * mm specified thickness | <br><br>nr<br>nr | | | |
| D | Excavate diaphragm wall panel to depth not<br>exceeding * m<br>    * mm specified diameter<br>    * mm specified diameter | <br><br>m²<br>m² | | | |
| E | Concrete quality * in diaphragm wall<br>    * mm specified thickness<br>    * mm specified thickness | <br>m²<br>m² | | | |
| F | Deformed grade 460 straight bars in reinforcing cage<br>    * mm diameter<br>    * mm diameter | <br>t<br>t | | | |

| Number | Item description | Unit | Quantity | Rate | Amount £ p |
|--------|-----------------|------|----------|------|------------|
| G | Plain grade 250 helical binding * mm diameter at * mm pitch | t | | | |
| H | Backfill open excavation with *, as specified in the particular specification | m³ | | | |
| J | Remove excavated material from the area around the diaphragm wall during boring operations, including loading and depositing * | m³ | | | |
| K | Cut down concrete panel to the specified cut-off level, prepare exposed head and reinforcement to receive capping: load and dispose of debris<br>    * mm specified thickness<br>    * mm specified thickness | <br><br><br>m²<br>m² | | | |
| L | Make concrete cube and test | nr | | | |
| M | Take bentonite sample from excavation and test | nr | | | |
| N | Remove obstruction in bore, utilizing standard equipment on site. Each plant and gang | h | | | |
| P | Charge for plant and labour delayed by causes beyond the piling contractor's control or when instructed. Each plant and gang | h | | | |

## Steel sheet piling

| Number | Item description | Unit | Quantity | Rate | Amount £ p |
|--------|-----------------|------|----------|------|------------|
| A | Transport steel sheet piling plant and equipment to site, set up, dismantle and remove upon completion. Per visit | sum | | | |
| B | Supply, handle, pitch and drive piles grade * self colour/painted PC * to * μ nominal D.F.T. or similar in lengths * m to * m<br><br>*Lengths (m) to be specified in the following ranges:*<br>    *2 to under 4*<br>    *4 to under 6*<br>    *6 to 18*<br>    *Over 18 to 24*<br>    *Over 24 to 29*<br><br>(Note: This is consistent with BS method of specification.) | m² | | | |
| C | E.O. for fabricated piles<br>(a) Open corner type 2A<br>(b) Closed corner type 2<br>(c) Junction type 3<br>(d) Junction type 1<br>(e) Other | <br>m<br>m<br>m<br>m<br>m | | | |
| D | Cut-off piles | m | | | |
| E | Track move plant and equipment in excess of 100 m | no. | | | |
| F | Derig move including low loader for rig | no. | | | |
| G | Re-transport steel sheet piling plant and equipment, set up, dismantle and remove on completion for pile extraction. Per visit | sum | | | |
| H | Extract piles as driven | m² | | | |
| J | Credit for goods, straight, reusable or unused piles in * minimum lengths<br>(a) * type | <br><br>ton | | | |

Although permanent works steel sheet piles would not be extracted, temporary steel piles are sometimes specified within the contract documentation if they are a necessary part of providing the permanent works.

*Specification for piling and embedded retaining walls.* Thomas Telford, London, 1996.

# Appendix B.
## Model clauses for incorporation as may be appropriate in main contracts or sub-contracts for piling works

**Inspection of site**

The Contractor shall be deemed to have visited and examined the site and its surroundings and to have satisfied himself, prior to the submission of his tender, as to the nature of the ground (so far as is practicable from visual inspection, and taking into account any relevant information which may have been provided in the tender documents), the form and nature of the site and its relationship to existing buildings both within and outside the site boundaries. He shall also have assessed the extent and nature of the labour and materials necessary for completion of the piling works, the means of communication with and access to the site, the accommodation which he may require, and generally to have considered all necessary information (subject to the above mentioned) as to risks, contingencies and all other circumstances influencing or affecting his tender.

**Ground conditions**

The tender shall be based on such factual ground investigation data and other information regarding the location, depth and condition of adjacent underground structures and services which may be affected by the piling work as shall have been provided within the tender documents.

In the event that ground conditions are encountered during the execution of the works which the Contractor considers to be more adverse than those which could reasonably have been foreseen at the tender stage, and which necessitate a change of design and/or method of construction, the Contractor shall immediately notify the Engineer and shall submit to the Engineer his proposals for overcoming such adverse conditions.

Following receipt of the Engineer's instructions, the Contractor shall give notice in writing, in accordance with the conditions of contract, of any intention on his part to submit a claim for extension of time and/or financial reimbursement. No such claims will be admitted if, in the opinion of the Engineer, the conditions actually encountered could have been reasonably foreseen at the time of tendering by a contractor experienced in piling works.

**Extent of contract**

The contract shall include the provision of all labour, materials, construction plant, temporary works, transport to, from, on or about the site, and everything required in/for the construction, completion and maintenance of the works insofar as the need for them is specified in or may reasonably be inferred from the contract documents.

**Additional requirements for main contracts**

Reference should be made to the items in Appendix C as a check-list on the drafting of contract documentation for piling as a main contract.

**Contract conditions**

Clause 42 of the ICE conditions (Possession of site) may require amplification for contracts under those conditions as follows: 'the allocation of working areas to the Contractor does not afford him exclusive rights of occupation; the Employer and other contractors shall have the right to complete freedom of access to all parts of the Works and to carry out any work that may be required in any place on the Site' but without disrupting the piling contractor's operations. Before including any such amplification for works carried out under other conditions it is desirable to verify whether or not it conflicts with the terms of that contract.

Additional clauses may also be required. For instance, in the case of piling for maritime work, the Contractor may require the right to use a berth for his craft and would have to know whether port and wharf dues were payable by him.

*Specification for piling and embedded retaining walls.* Thomas Telford, London, 1996.

# Appendix C.
## Schedule of attendances and facilities to be provided by the client to a Steel Piling Specialist

**FEDERATION OF PILING SPECIALISTS**

COMPANY NAME: .....................................................

CONTRACT NAME: .................................................

TENDER REF: .....................................................

**FPS**

### SCHEDULE OF ATTENDANCES AND FACILITIES TO BE PROVIDED BY THE CLIENT TO A STEEL PILING SPECIALIST

For the purposes of this document the following definitions shall apply:

| | | |
|---|---|---|
| Specialist | — | Steel Piling Specialist |
| Client | — | Person directly employing the Specialist |

The following attendances and facilities shall be provided and maintained at all times (including additional working hours if necessary) for the duration of and in relation to the specialist works, free of charge to the specialist and in a manner so as not to disrupt or restrict the regular progress of the specialist works.

1. **Notices.** Giving all notices and obtaining all necessary approvals, licences and sanctions, including but not limited to any wayleaves, easements, possessions, rights of way or access.

2. **Rates and Fees.** Payment of any rates or fees which may become payable due to occupation of the specialist works.

3. **Protection.** Protection of the works where taken over by other traders or contractors or where the specialist has left site.

4. **Watching.** Provision of security to safeguard the plant, equipment, materials on the site and the specialist works.

5. **Fencing, Hoardings, etc.** Hoardings, fences, noise and splash barriers, statutory warnings, flagmen or the like as necessary to protect the works, plant, materials, personnel, third party property, and the general public. This shall include protection from exhaust, oil, grease, etc.

6. **Clearance.** The provision of adequate clearance around working positions for the specialist's operations including protection to adjacent works and third party property.

7. **Access and Hardstandings.** Full and free access on to the site(s), including adequate means of access from hard road to firm dry level all-weather working surfaces, with adequate working space including staging/protective mats where necessary, prepared and maintained in a manner suitable for the safe movement on to and off the site(s) and to and between working areas, storage areas, pile/panel positions and test piles/panels for mobile plant and equipment and wheeled transport including articulated lorries to within 6 m of piles to be driven and 4 m of piles to be extracted. Ramps, including access ramps storage areas where required, to a gradient not steeper than 1 in 10. Protective mats and all other equipment and measures necessary to minimise the risk of damage to third party property, including road surfaces and kerbs especially where the nature of the site requires off site activity such as loading and off-loading of materials and plant which might include crawler cranes.

8. **Surface Water and Groundwater.** Any pumping dewatering or drainage required to keep the site and works free of surface water or any water arising from the operations.

COMPANY NAME: ...............................................

9.  **Accommodation and Storage.** Provision and subsequent removal if required of adequate firm, dry, reasonably level working areas, prepared and maintained in a manner suitable for the safe operation and erection of plant and equipment. Conveniently situated areas on site for storage of plant, equipment and materials, for offices, sheds and the manufacture of corner piles and the like, prepared and maintained as per the access routes.

10. **Flammable Stores.** Provision for storage of petroleum, explosives and flammable materials as may be required and arranging for the requisite licence.

11. **Telephone Facilities.** Provision of site telephone facilities.

12. **Health and Safety.** Welfare and safety facilities to comply with statutory regulations or rules, orders or regulations of any authority having powers related to the specialist works. In operations over or near water, provision and maintenance of proper and efficient life saving equipment at all times including safety boats, netting, and life belts .

13. **Temporary Lighting.** Suitable background and task lighting to working areas to allow safe working and safe access and egress and to facilitate execution of the specialist works.

14. **Water Supply.** Within the working, storage and preparatory operation areas, potable water supply at mains pressure take-off points and sufficient for the operations.

15. **Electricity.** Within the working, storage and preparatory operation areas, suitable power take-off points and power.

16. **Traffic.** Control or diversion of road, rail or water-borne traffic.

17. **Existing Services.** Clear and substantive setting out, marking or exposing on site the exact location of existing underground/overhead works and services and providing a drawing on which their positions in line and level are accurately plotted relative to the specialist works. Adequate protection, diversion or removal of such works or services to prevent damage from the specialist's operations.

18. **Shoring/Underpinning.** Shoring and underpinning as necessary, including the removal, replacement or adjustment of timbering or shoring which may impede the specialist's operations.

19. **Leader Trench.** Excavation and subsequent backfilling of leader trench to the specialist's requirements.

20. **Obstructions.** Prior removal of overhead, surface or underground obstructions (whether causing refusal or deflection of piles) which may, in the opinion of the specialists, impede the penetration of the piles to a depth or tolerance specified or requested, or the installation of framing materials. Backfilling of excavations and voids with a suitable material which will not obstruct or be deleterious to the works but which will ensure the stability of the specialist's plant.

21. **Setting Out.** Clear and substantive setting out and maintenance of pile positions as necessary throughout the contract and the provision of permanent datum points, base lines and structural grid lines.

22. **Checking.** Checking the positions and cut-off levels of all piles/panels, during the progress of the work, on completion of the work (where practicable) and before the specialist's plant has left site.

23. **Removal of Material.** Removal and disposal of excavated or displaced material in sufficient time to prevent the formation of spoil heaps impeding the specialist's operations.

24. **Wheel and Road Cleaning.** Manned wheel-cleaning facilities and/or road-cleaning, as necessary.

25. **Trimming.** Pile tops will be burnt off when instructed to within 25 mm of a given level and maintained by client.

# Appendix D.
## Guidance on static load testing of piles

Axial compression load testing may be carried out by

(*a*)   maintained load testing (ML test)
(*b*)   the constant rate of penetration test (CRP test).

It may be desirable in unfamiliar ground conditions to test at least one preliminary pile for each major grouping of piles on a site or at least one preliminary pile for every particular set of ground conditions into which piles will be installed. In deciding the number of contract piles to be tested it will be necessary to consider

(*a*)   the requirements of the local statutory authority
(*b*)   the extent and reliability of the ground investigation and variations in ground conditions actually encountered
(*c*)   knowledge of the behaviour of similar piles in similar circumstances on adjacent sites
(*d*)   the standard of workmanship and of supervision on the site
(*e*)   any unusual difficulty or major variation encountered in the formation of the contract piles
(*f*)   the extent of integrity-testing being carried out.

The measurement of the movement of the pile head should be carried out by one of the methods described in the *Specification for piling and embedded retaining walls*. However, it should be noted that the establishment of a reference frame on shallow supports which is unaffected by extraneous movements is often difficult. Movement may be caused by any of the following

(*a*)   ground displacements due to changes of loading in the vicinity: for example, as the kentledge load is removed from the ground surface during a test, the vertical displacement of a reference frame supported on shallow foundations on soft clay can be 20 mm or more
(*b*)   the effects of temperature and wind on the datum frame structure: the direct rays of the sun cause differential movements in the frame and wind causes vibration; a frame which is shielded from the sun and wind by boxing will still be subject to daily variation of temperature, which may cause a cyclic displacement of 0.25–0.5 mm
(*c*)   construction plant operating in close proximity to test equipment.

Cyclic movements and their pattern due to causes as in (*b*) can be established if a series of readings is taken before the loading test starts. The effects of cyclic movements may be minimized in a test of long duration if readings are taken at the same time each day. In general, the results of CRP tests are little affected by the foregoing influences.

When approving the reference frame and its foundations, the Engineer must consider the type of test and the precision required

---

in the measurement of movements. An improvement in the accuracy may be obtained if the reference frame is supported on piles, or if deep datum points are formed at each test location in order to measure by levelling methods the movement of the frame relative to a stable datum. If the precision required in the maintained load test is greater than that obtained with the levelling method normally available to the Engineer, then an optical level incorporating a parallel plate micrometer should be used. This enables an accuracy of 0.1 mm to be obtained.

The Engineer should state in the contract documents the required maximum load to be applied to each test pile and the Contractor should provide the necessary reaction. When deciding the magnitude of this reaction, account should be taken of the possible variations in the soil strata, which may not have been revealed in the ground investigation and which may result in different ultimate capacities from those estimated. Provided the structural integrity or stability of the preliminary test pile is not likely to be damaged and the cost of a pile test is within economic and practical limits, consideration should be given to making the maximum test load of such magnitude that soil failure will occur. Such a test load will give a guide to the real factors of safety and may assist in the interpretation of variations in the characteristics of the contract piles. Contract piles should be subjected to proof load tests, commonly the design verification load plus 50% of the specified working load (see clause 10.13 of *Specification for piling and embedded retaining walls*).

In a CRP test on an end-bearing preliminary pile, difficulty is sometimes experienced in identifying the point on the force-penetration diagram at which ultimate bearing capacity is reached. Ultimate bearing capacity of the base is normally not reached until the penetration is at least equal to 10% of the base diameter. However, the value for unplugged steel piles may require only two to three times the wall thickness, i.e. 20–40 mm. Much greater penetration may be necessary to achieve rupture load, i.e. change in settlement for change in load is infinite. At smaller penetrations only a portion of the ultimate bearing capacity is mobilized. Often a penetration of the pile head equal to 20% or more of the base diameter is required in frictional soils to mobilize the full bearing capacity of the pile. For unplugged steel piles, this may be only 40 mm. The minimum penetration to be achieved should be specified and an Engineer wishing to make a closer approach to the value of ultimate bearing capacity by means of the CRP test should consult the literature (Whitaker, 1963; Whitaker and Cooke, 1961; England and Fleming, 1994). The Engineer should specify the maximum penetration required and the Contractor should be required to provide the requisite reactions, equipment and measuring devices.

The rate of penetration to be used in a particular test should be chosen with regard to the nature of support to the pile, the elastic deformation of the pile, the time required to take and record readings and the number of readings required to plot an unambiguous curve of load against movement. The pump supplying the hydraulic jack must be capable of maintaining the specified rate. The movement of the base of the pile will be smaller than that of the head because of the elastic deformation of the pile under load. The elastic deformation may be estimated or may be measured.

If the pile is end-bearing on rock or on very dense sand or gravel, a short stiff pile may reach the maximum specified test load with a very small penetration of the base and consequently little downward movement of the head. A long pile, although having a small base penetration, may show an appreciable downward movement of the head due to elastic compression. An estimate must be made of this downward movement so that sufficient readings may be obtained to cover the range from zero to the maximum test load.

If the pile is end-bearing in medium dense sand or gravel, a rate of penetration of 0.02 mm/s is generally convenient for piles up to 20 m long. For a very long pile this rate may be increased to allow for the elastic compression of the pile.

If the pile obtains the greater part of its support by friction on the pile shaft, the movement to reach ultimate bearing capacity may not exceed 1% of the shaft diameter. A convenient rate of penetration for most friction piles is 0.01 mm/s.

It is advisable to specify that a plot of load against penetration is to be made as the tests proceeds so that the appropriate point at which ultimate bearing capacity is reached can be identified and the test terminated after an appropriate penetration has been achieved.

It should be borne in mind that the CRP test is not designed as a method of determining pile settlement at a given load. For this reason, a composite test is commonly used and envisaged in the *Specification for piling and embedded retaining walls*, entailing maintained test-loading to a prescribed figure, followed by a normal CRP test.

Lateral and axial tensile load testing of piles is undertaken infrequently and no standard procedures have been evolved for these tests. For information on procedures and interpretation, reference may be made to the literature (de Beer, 1977; Broms, 1981; Patel, 1992). Care should be taken to ensure that the application of load for such a test reflects the manner in which the pile will be loaded in the permanent works.

# Appendix E.
## Guidance on dynamic load testing of piles

Piles may be tested by impulse loading using a piling hammer to apply a test load to a pile. This test was developed initially for driven preformed concrete, steel or timber piles. The method has, however, been applied to uncased and cased in-situ concrete piles.

Dynamic pile-testing is normally used to evaluate the pile capacity, soil resistance distribution, immediate settlement characteristics, hammer transfer energy (efficiency), and pile stresses during driving. The results obtained relate directly to dynamic loading conditions.

The test results can be applied to

(*a*) evaluate and verify pile design at preliminary or subsequent stages of work
(*b*) proof-test certain piles as work proceeds
(*c*) evaluate pile stresses during installation (to avoid pile damage)
(*d*) evaluate hammer energies to validate input data for driving formula
(*e*) evaluate parameters to verify wave equation input data
(*f*) evaluate assumptions made of pile driveability.

Test piles are instrumented with strain transducers and accelerometers within two pile diameters of the top of a pile. Measurements of strain are made under hammer impact and, at the same time, the motion of the pile is measured either as acceleration or as displacement. The test data from each hammer blow, or from selected hammer blows, is recorded for further analysis.

It is fundamental to dynamic pile-testing that the soil resistance to pile motion is considered to be both static (elasto-plastic) and dynamic (damped). Methods have been devised to evaluate static resistance during testing, but they rely on an assumption of the soil damping resistance and should normally only be used where the damping resistance has been evaluated and validated by other means (e.g. static load testing a pile).

Dynamic pile-testing and particularly the analysis of test data must be carried out by trained and experienced engineers. Generally, it is advisable to subject the data obtained from dynamic tests to subsequent rigorous analyses which are normally carried out remote from the piling site. Such analyses are normally based on wave equation philosophy and entail some form of computer program. Rigorous analyses give more detailed results than those obtained from direct on-site evaluation and include pile bearing resistance distribution and inferred immediate pile deformation characteristics under static load. It is important to note that, since the dynamic test load is applied for a very short duration, inferred settlement under static load does not include any long-term effects such as soil consolidation and creep. Thus, although dynamic pile-testing will provide a rather comprehensive insight into pile–soil interaction, long-term settlements will not be predicted and appropriate soil mechanics methods of calculation should be used. Where ultimate bearing capacity evaluation may be desirable, it will not always be obtained readily owing to pile 'refusal' under the test hammer blow. Too little pile movement to mobilize the full soil resistance forces may have occurred. A more powerful hammer may be needed for testing purposes, but the consequences of overstress in the pile and additional cost should be assessed.

Where preliminary pile-testing is not considered justified, a greater number of dynamic proof tests on working piles may be carried out than has hitherto been justifiable economically. Consideration should be given to the combination of static and dynamic load tests, having regard to the Engineer's experience of dynamic load testing, and his knowledge and experience of the ground conditions. These tests would increase confidence in the design or enable it to be modified prior to commencement of the installation of working piles.

The time which should elapse before piles are tested needs to be considered carefully. Ideally this should have regard to soil characteristics and pile installation processes, but practical and economic factors will influence decisions and test information requirements. Where it is required to evaluate pile driving stresses, testing must be carried out during the installation procedure. Where bearing and resistance evaluation is required, it is desirable that testing should be carried out some time after installations, as in the case of static load testing.

# Appendix F.
## Guidance on integrity testing of piles

The traditional methods of exposure of pile shafts by excavation, coring, etc., are still in common use, particularly as a means of verifying the interpretation of modern indirect methods of integrity-testing.

The position of a borehole drilled in a pile shaft relative to the pile axis may become uncertain as the depth from the surface increases, so that accurate location of any defect within the pile cross-section is unlikely.

If the borehole is formed by coring, then inspection of the core extracted will give an indication of the concrete quality. The borehole only enables identification of the material which is brought out or which may be observed at the sides of the hole. The nature of the remainder of the pile shaft has to be deduced from the circumstances.

A significant advance in identifying the existence of defects in the construction of piles has been the development and adoption of modern integrity-testing systems. Methods which are used to investigate the integrity of concrete piles are as follows.

(a) *Time-based techniques: echo testing.* These methods have been developed independently by the Central National Council for Applied Scientific Research (TNO) in Holland, and by the Centre Expérimental de Recherches et d'Etudes du Bâtiment et des Travaux Publics (CEBTP) in France. The pile head is impacted by a relatively small hammer, and the response is measured and processed in such a way that most major discontinuities should be apparent.

(b) *Frequency-response testing.* The steady-state method has been developed in France by CEBTP and has achieved widespread use. Basically it involves attaching an electrodynamic vibrator to the pile head and measuring the response of the pile to a sinusoidal force of constant amplitude. By varying the frequency of vibration, resonance peaks are produced, which may be used to interpret the effective length of the pile and to display most major defects. The stiffness of the head of the pile can also be measured by this technique. A transient frequency-response method has been developed by Ellway.

(c) *Cross-hole sonic logging.* This requires that tubes be cast into the pile to allow the passage of a sonic pulse from a transmitter to a receiver through the material of the pile. The usual procedure is to measure the elapsed time between the sending and receiving of a signal between the tubes. The costs of tube installation are high, but the method has nevertheless proved to be both practical and useful.

An advantage of the time and frequency-response methods of testing lies in the fact that special provision, such as the installation of tubes, does not have to be made during construction of a pile and that testing can be carried out quickly and without significant disruption to normal site activities. The costs are small by comparison with load-testing, coring or excavation, and a large sample of the piles, or perhaps all of the piles, on any site can be examined for potential defects.

It must be emphasized that the results of integrity-testing need to be interpreted by engineers with the requisite specialist experience, and that all methods have limitations. It cannot be construed that methods of integrity-testing will identify all imperfections but they provide a useful tool, particularly as a safeguard against major faults existing within the effective length evaluated by a particular test. It should be appreciated that anomalous results can arise which may be capable of alternative interpretations. Integrity-testing may also identify minor defects which will not necessarily affect pile performance, and the Engineer will have to exercise his judgement as to the acceptability or otherwise of such a pile.

Useful references descriptive of the methods and their application are given in a CIRIA review (Weltman, 1977; Turner, 1966). Many other references exist and a summary of the findings from the TNO method in recent usage is given by Sliwinski and Fleming (1984).

*Specification for piling and embedded retaining walls.* Thomas Telford, London, 1996.

157

# Appendix G.
## Drafting of a performance specification

The procedure referred to in paragraph 2.1. may be operated in various ways. The two most common applications are as follows.

(a) The Engineer provides a piling layout showing the locations of individual piles required to carry stipulated working loads, and may specify a minimum cross-sectional area, but does not specify the length of the individual piles.

(b) The Engineer provides a plan showing the amount and location of vertical loads and any other forces to be carried by the piled foundation but does not specify the working loads of the piles. In this case the specialist piling contractor is responsible for determining the required layout of the piles as well as their individual design and should take into account any local eccentricities which may arise due to tolerances in position of the piles.

The attention of the Contractor should be drawn to special loading conditions caused by surcharging, removal of support, downdrag or the like.

In order that the Engineer may ensure that the piled foundation and the structure to be supported will be compatible in terms of soil–structure interaction, he must consider the behaviour not only of an individual pile but also of the groups and the overall piled foundation before specifying the performance requirements for the piles (Institution of Structural Engineers, 1988).

Where a piling layout is provided by the Engineer, the specified working loads to be carried by the individual piles should be stated, together with the limitations on settlement at the head of each individual pile type when subject to a specified proof load, and having due regard to the subsoil conditions. It follows that these criteria can be specified only by the Engineer responsible for the design of the structure as a whole, who is in a position to form a judgement as to the capacity of the structure to resist total but more importantly differential settlement. He must therefore ensure that the performance criteria take into consideration the group action of all the piles supporting a structure.

When the working load of the individual piles is not specified, the specification of performance requirements becomes more difficult and criteria have to be related to the support of the individual loads which may be carried by individual piles or by pile groups, in terms of both factors of safety and relative settlements at service loads.

Irrespective of the type of performance specification, the Engineer must avoid requirements which are unrealistic in terms of the soil conditions within which the piles are to be constructed. He must take account of the fact that any limitations on settlement of the head of the pile must be greater than the elastic response of the pile shaft, having due regard to load-shedding through shaft adhesion.

Although the use of a performance specification may impose a design responsibility on the specialist piling contractor, subject to the terms and conditions of the main contract, the Engineer must bear in mind that this may not relieve him of his own ultimate design responsibility for such work and, in particular, the duty to satisfy himself that the type and design of the piles offered by the specialist contractor are suitable for construction in the ground conditions and are compatible with the site environment.

The Contractor should be required to furnish with his tender full details of the type of pile offered, the standards of control he intends to use, how the calculation and checking of the load-bearing capacity and settlement of the piles will be carried out, and the tests he proposes to undertake on site.

# References

Broms B.B. *Precast piling practice.* Thomas Telford, London, 1981.

de Beer E. The effects of horizontal loads on piles due to surcharge or seismic effects. *Proc. 9th Int. Conf. Soil Mech., Tokyo,* 1977, **3,** 547–558.

England M. and Fleming W.G.K. Review of foundation testing methods and procedures. *Proc. Instn. Civ. Engrs, Geotech. Engng,* 1994, **107,** July, 135–142.

Institution of Structural Engineers *et al. Soil Interaction.* ISE, London, 1988, 2nd edn.

Patel D.C. Interpretation of results of pile tests in London Clay. *Piling Europe Conference.* Thomas Telford, London, 1992, 91–101.

Sliwinski Z.J. and Fleming W.G.K. The integrity and performance of bored piles. *Piling and ground treatment.* Thomas Telford, London, 1984, 211–223.

Turner M.J. *The role of integrity and other non-destructive testing in the evaluation of piled foundations.* Construction Industry Research and Information Research Association, London, 1996.

Weltman A.J. *Integrity testing of piles: a review.* Construction Industry Research and Information Association, London, 1977, report PG4.

Whitaker T. The constant rate of penetration test for the determination of the ultimate bearing capacity of a pile. *Proc. Instn Civ. Engrs,* 1963, **26,** Sept., 119–123.

Whitaker T. and Cooke R.W. A new approach to pile testing. *Proc. 5th Int. Conf. Soil Mech., Paris,* 1961, **II,** 171–176.

# Guidance notes

# Section 1.    General requirements for piling work

1. Sections 1 to 10 and Section 20 of this Specification are based on the Institution of Civil Engineers (ICE) *Specification for piling* (1988) with substantial modification. The requirements of this Specification are very similar to the Highways Agency *Specification for highway works*, Series 1600, August 1994 issue.

   The 1988 ICE *Specification for piling* has been modified to remove clauses which contain guidance rather than specification and these guidance clauses have been incorporated in the following sections of guidance notes. Reference in the Specification to approval by the Engineer has also been removed from the text, as this is normally covered by the Preliminaries and the Conditions of Contract. The Contractor is required to submit details of his proposals in accordance with the schedule of submissions in Table 1.1. Variations and instructions to the specified requirements or the submitted details should be dealt with under the provision of the Contract by Variation Orders where appropriate.

2. The document has been assembled so that the specifier selects the clauses of the specification considered to be appropriate for the Works and completes the appropriate Particular Specification for each clause selected. Section 1 will be required in all cases. Section 10 will be required in all cases except for Engineer-designed piles where the Engineer has decided that there are sufficient load test data in the vicinity for the behaviour of the chosen pile in the particular ground conditions to be well understood. Section 20 will be required for all piles made of reinforced concrete. The other sections will be selected depending on the particular requirements of the structure, site and environmental constraints. In many cases there will be more than one type of pile that could provide an appropriate solution and the Engineer may choose to allow the Contractor to select from a range of possible pile types. A list of the requirements to be specified is given in the Particular Specification. Other requirements can also be specified by adding sections to the Particular Specification.

3. Options are provided for either the Engineer or the Contractor to design the piles. Guidance on responsibility for design is given elsewhere in this Volume. The Engineer can either specify that the Contractor designs the pile (Option 1) or he can provide his own design (Option 2). If Option 1 is selected then the Engineer must provide performance criteria for the piles and sufficient data on the loads and moments to be applied to the individual piles and the overall performance requirements for the structure for the Contractor to be able to assess the total requirements for the piles. Further guidance on the criteria to be provided for the structure is given in the Particular Specification and loading and performance criteria

for the various pile types are included in the Particular Specification for the relevant pile type. For an Option 1 design the Engineer should obtain from the Contractor full details of the type of pile offered, the standards of control he intends to use, how the calculation and checking of the load-carrying capacity and settlement of the piles will be carried out and the proof loading or other testing he proposes to undertake on the site. Where the Engineer prepares a design (Option 2), the piling type he has selected must be based upon the use of a non-proprietary system. The following information should be shown on the Drawings (cross-referenced in the Particular Specification):

- pile layout
- working loads
- location of preliminary piles.

4. A comprehensive site investigation carried out in accordance with BS 5930 is an essential prerequisite to the selection, design and construction of piles. Particular attention shall be given to the following aspects, which are particularly relevant to the construction of piles:

- a comprehensive desk study of the site history to assess the risk of obstructions, contamination, quarries, opencast and deep mines, backfilled sites or archaeological finds that could affect the feasibility of pile construction, programme and cost; suspected obstructions should be identified by probing or preliminary enabling works to confirm their extent;
- equilibrium piezometric levels of all possible water tables, including artesian conditions and any seasonal or tidal fluctuations;
- permeability of the soils;
- presence of coarse, open soils, cavities, natural or artificial, which may cause sudden losses of support fluid in open excavation and require preliminary treatment;
- strength and deformation characteristics of soils, particularly weak strata that could cause instability or large deformation;
- the presence of boulders or obstructions which may cause difficulties in excavations or driving conditions;
- soil and groundwater chemistry which may affect the durability of the piles, the disposal of spoil, and the performance or disposal of support fluid;
- the strength and profile of any rock surface beneath the area to be piled.

Conditions of Contract may affect what information should be provided in ground investigation reports and the responsibility attracted by the various parties particularly where interpretation is included in the reports. Reports which should include both factual and interpretative ground investigation reports given to the Contractor should be checked to ensure they are relevant to the structure to be built and that the scope of construction works envisaged at the time the site investigation was carried out have not significantly changed. If they have or if the ground investigation reveals additional problems, further investigation should be carried out.

5. Unexpected and emergency situations may arise which, if not dealt with rapidly, could jeopardize the integrity or performance of the completed foundation. It is essential for foundation construction that the contractual arrangements between the various parties accept the possibility that unexpected conditions can occur and that remedial action can be agreed sufficiently quickly for the work to continue without any adverse effects on the completed foundation. For this reason full time supervision by the Engineer and the Contractor is recommended for all piling works. These site representatives should be sufficiently familiar with the construction of the piles and the design requirements to enable solutions to unexpected or emergency situations to be agreed without unreasonable delay.

6. The tolerances specified are realistic for most sites and ground conditions. However, where the ground contains obstructions or for certain ground conditions (e.g. pile tip moving from soft to dense layers), tolerances may need to be relaxed. This will necessitate an allowance in the design of pile caps and ground beams to suit the installed pile positions. The Contractor should be informed of the reasons leading to the changed tolerances. The Engineer should ensure that the design of the structure to be built on the piles is not compromised. In some cases the design may require tighter tolerances than those specified in which case the required tolerances should be specified in the relevant Particular Specification. Plunged stanchions may need to be positioned to structural tolerances which are much tighter than those for piling. If this is the case the tolerances should be specified in the relevant Particular Specification and the Contractor's method statement should show how the tighter tolerances are to be attained. Casting tolerances of bored piles above cut-off level are specified in Table 3.2. The Contractor may overpour the piles to whatever level within the tolerance that he wishes but he must ensure that concrete at cut-off level is dense and homogeneous. The bill of quantities should allow for the cutting down and removal of overpoured piles to the tolerance set down for the particular conditions. If the Contractor cannot comply with the specified tolerances, he should provide details with his tender. Any additional costs due to his non-compliance should be taken into account in the tender comparison.

7. The piling method statement should include sufficient detail to demonstrate that it is compatible with the design assumptions and any site constraints. The amount of detail required at the time of tender should be considered, bearing in mind what is required for an initial assessment and comparison of tender proposals and what is reasonable for a Contractor to provide in a competitive situation. The Contractor's designs must be regarded as confidential (copyright laws apply) and parties receiving such designs should, if the Contractor so requests, be prepared to sign confidentiality agreements. The following list includes examples of items that may need to be provided:

- site staff and organization;
- the experience of the Contractor and his staff with the piling type and the particular ground conditions;
- the details of piling plant should be sufficient to demonstrate its suitability for achieving the required penetration and to work within any noise and vibration limits where these are restricted;
- the sequence of piling which may affect uplift and lateral displacement of driven piles or affect the integrity of nearby cast-in-situ piles;
- setting out and means of achieving specified tolerances;
- the time period for boring and concreting that may have an influence on the design assumptions for the pile resistance that can be mobilized;
- the length of temporary casing or use of support fluids to maintain adequate support to the ground during construction;
- the method and equipment for cleaning or forming the base of the pile where end bearing is assumed in the design. The design assumptions for end bearing should take into account what can realistically be achieved with the proposed method;
- the methods that will be used to check pile depth, bore stability, base cleanliness and compliance with the specified tolerances;
- the method of placing concrete to ensure it is placed vertically down the bore. Where concrete is placed under water or support fluid the method of placing concrete by tremie should be described;
- the frequency and means of testing concrete workability, strength and compliance values for grouts;
- the means of placing reinforcement including the lapping of reinforcement cages, the method of maintaining concrete cover and the vertical position of the cage and any additional reinforcement for lifting, handling and placing;
- details of permanent casings and how they are to be maintained in position during concreting;
- details of preboring, or other means of aiding pile driveability and measures that will be taken to minimize disturbance of the surrounding ground;
- the level to which concrete will be cast and details of how completed piles are to be protected from subsequent damage;
- procedures for dealing with emergency situations such as sudden loss of support fluid, obstructions and piles that are out of tolerance;
- company safety plan;
- quality assurance procedures;
- typical record sheets.

For Contractor designed piles or for Contractor proposed alternatives, the Contractor must gain approval for his design from the relevant authorities.

8.  The Engineer will normally determine the location and condition of adjacent structures and services that are likely to be within the zone of influence of the Works. Certain structures may be particularly susceptible to noise, vibration

or ground movement and the Engineer should assess prior to inviting tenders where special measures are likely to be required for protecting these structures. The information on location and condition of these structures together with the restrictions to be imposed and the monitoring requirements should be presented by the Engineer on the Drawings and in the Particular Specification. The Contractor is required to confirm the information on site and to provide proposals to meet the requirements of the Engineer in respect of these structures and services.

9. Should an excavation be made alongside completed piles they will be subjected to lateral loading and the design should allow for such conditions where necessary.

10. British Standard 8004 : *Code of Practice for foundations* provides guidance on preliminary investigation, design, materials, safety, workmanship and control of piling.

11. The following notes should be followed when pile performance is being specified in Table 1.3. Each note corresponds to a column in the Table, reading from left to right:

   - Each and every pile should be allocated a unique reference number or code;
   - Permitted types of pile are restricted to those specified in the corresponding Section of the Specification, e.g. insertion of 'Section 3' will restrict permitted type(s) of pile to bored cast-in-place piles only. Further specification, e.g. 'underreams not permitted' may be necessary in particular circumstances;
   - Working Loads specified on drawings should be grouped and each group allocated to one Allowable Pile Capacity, e.g. all piles with Specified Working Loads between 858 kN and 930 kN are to be constructed for an Allowable Load of 930 kN. 'Grouping' of Working Loads in this way reduces the number of different pile sizes on site and helps eliminate the confusion that can arise when each pile is individually sized;
   - The pile designation relates to the size of the particular pile and possibly to the reinforcement cage. The number of different pile sizes for the project will then be listed separately.
   - The DVL may be much larger than the Allowable Pile Capacity;
   - The Load Factor should be specified;
   - A realistic estimate of the likely lower bound load settlement curve for a pile tested in isolation should be made, and the Permitted Settlement at 1 DVL taken from that curve. If a stiffer pile than is indicated by the Permitted Settlement is required, then the pile type or dimensions will have to be changed (e.g. pile lengthened, larger diameter, or underream added);
   - The stratigraphy of the ground may make it imperative that piles have a minimum length, to ensure penetration into a particular stratum; alternatively the minimum penetration into a particular stratum can be specified (the column

heading will then require to be changed to 'Minimum penetration into XYZ');

- The minimum pile diameter will normally be determined by the permitted stresses in the pile materials, taking account of axial loads, moments and transverse loads.

12. If it is not the Engineer, the body (Architect, Contract Administrator, etc.) acting as supervising officer should be specified in the Particular Specification. If that body's powers are not clearly defined elsewhere, then details should be provided there.

13. Guidance on the control of noise and vibration due to piling operations is given in BS 5228: Part 4.

# Section 2. Precast reinforced and prestressed concrete piles and precast reinforced concrete segmental piles

1. This section applies to precast reinforced and prestressed concrete piles usually supplied for use in a single length without facility for joining lengths together and to piles made of precast reinforced concrete elements cast at a precasting works away from the site, where work cannot normally be closely supervised by the Engineer. The segmental elements are joined together as necessary on site during driving, using special proven steel joints, incorporated into the pile elements when cast.

2. Sub-contracting of the manufacture of precast reinforced and prestressed concrete piles requires the Engineer's written consent under Clause 4 of the ICE Conditions of Contract, 6th Edition. If their manufacture differs in any respect from that specified in the Particular Specification, the Contractor should be asked for complete details of his proposed alternative if these were not submitted at tender stage so that the proposed alternative can be evaluated.

3. Cement, aggregates and water shall comply with Section 20.

4. The minimum concrete grade for precast reinforced piles is C30. The minimum cement content and maximum free water/cement ratio for various exposure conditions are given in Table 3.4 of BS 8110 where concrete in non-aggressive soil (e.g. Class 1 sulphate conditions) is defined as a moderate environment. Piles for highway structures should be designed in accordance with Table 13 of BS 5400: Part 4. In this standard, buried concrete is classified as being subject to severe exposure conditions or moderate for the case of concrete permanently saturated by water with a pH greater than 4.5. These conditions are more onerous than BS 8110. Nominal cover to reinforcement for Grade C30 concrete for buried parts of the structure is 45 mm. For hard driving conditions or where reduced nominal cover of 35 mm is required, Grade C40 concrete is recommended with a minimum cement content of $400 \, \text{kg/m}^3$ according to BS 8004. The greater the cover, the greater the tendency for concrete to spall off during hard driving. For prestressed concrete piles the recommended minimum concrete grade is C40 with a minimum cement content of $400 \, \text{kg/m}^3$. The cement type, minimum cement content and maximum free water/cement ratio shall be in accordance with BRE Digest 363 to protect buried concrete from acid and sulphate attack.

5. The Engineer should require the Contractor to submit his detailed proposals for pile driving in compliance with Clause 14 of the ICE Conditions of Contract, 6th Edition (or other Conditions of Contract as appropriate). The designer should state the particular requirements in the Particular

Specification for the minimum length or set to be obtained. Generally in cohesive soils where piles are designed to carry the load in friction they should be driven to a specified penetration or length. Piles that are predominantly end bearing in cohesionless soils or founded on rock are driven to a set which in some instances is also combined with a minimum length requirement. Guidance on hammer selection is given by Tomlinson (1995). Hammers that are too light or whose delivered energy can deteriorate in energy output can lead to false sets. Effective hammer energy can be measured by dynamic testing. Normally, for a well-matched hammer and pile system the set will be between 10 and 25 blows for 25 mm penetration. The set for working piles should be established from experience or from a successful preliminary pile test. Set calculations are often not reliable.

6. The length of pile required should be specified by the designer and may be subject to the results of preliminary pile tests. Trial drives are recommended prior to preliminary pile tests or installation of working piles to assess likely variations in driving conditions across the site, to confirm the required sets, and the potential for uplift or lateral displacement of piles driven in groups. The effect of uplift on piles already driven is to reduce their end-bearing capacity and will become a critical factor for the control of piling when this is a main source of capacity, as is often the case for driven piles. Lateral displacement may also cause damage to piles already driven or adjacent structures. Careful trials at the commencement of piling can be used to determine the criteria for pile installation. Sometimes for a preliminary pile test, the test pile is driven first followed by its surrounding anchor piles in an uplift trial. The test pile can then be subjected to a static load test and if successful new permissible settlement and/or uplift criteria can be set. To minimize uplift and lateral displacement, piles are usually driven in order from the centre of a pile group outwards or away from an adjacent structure. Hammond *et al.* (1980) present data on the uplift of piles due to the driving of adjacent piles for particular pile types and ground conditions. Onerous settlement criteria may not be compatible with the permissible uplift criterion specified in Sections 2, 5, 6 and 7.

7. Reboring may be used to ease pile driving through dense layers or can be used to reduce lateral movements of the surrounding ground. It may affect the capacity of the pile. Jetting can be used to assist pile driving in certain cohesionless soils but it must be used with great care as it may affect the pile capacity and may also wash away soil supporting adjacent structures.

8. The Specification calls for full records to be made for the driving of every pile. Where consistent driving conditions have been established across a site the Engineer may choose to relax this requirement. As a minimum the Contractor should make a full record for the first pile in each area and for the final 3 m of every pile. In addition, full driving records should be made for at least 5% of the piles driven.

9.  A feature of driven piles in cohesive deposits is that as the soil is sheared during pile installation, the soil surrounding the pile loses strength. However, once the pile has been driven the soil consolidates and 'sets up' around the pile, giving increased capacity. Conversely, in fine granular soils (such as silt), dilation of the soil can cause negative pore water pressures (suction), which increase the driving resistance, so giving a false set. Dissipation of the suctions permits the soil to 'relax', so giving a reduced capacity. The timing of dynamic and static pile testing should take this into account.

10. Joints for segmental piles are generally made of steel and are therefore susceptible to corrosion if exposed to free oxygen and water. Consideration should be given to the location of the joints in completed piles. Generally in relatively low permeability soil beneath the water-table there is a low risk of joint corrosion. The risk increases if the joints are located in flowing groundwater conditions in permeable strata at or close to the water-table. Special care should be taken if joints or other steel elements are to be located in contaminated soils where anaerobic bacteria can attack the steel. Dock mud is a good example of such a zone. A problem with the final position of a joint can arise when pile driving reaches refusal before the pile is driven to its intended depth.

11. Heavy mechanical breakers to cut down piles should be used with caution as they may induce damage below the point of application. This applies particularly to piles of less than 600 mm diameter.

# Section 3.  Bored cast-in-place piles

1. This section applies to bored piles in which the pile bore is excavated by rotary and/or percussive means using augers, buckets, grabs or other boring tools to advance where possible a stable open hole. Where the bore is unstable, temporary or permanent casing or support fluids may be used to maintain the stability of the bore during excavation and concreting.

2. Cement, aggregates and water shall comply with Section 20.

3. The minimum concrete grade for bored cast-in-place piles is C30. The minimum cement content and maximum free water/cement ratio for various exposure conditions are given in Table 3.4 of BS 8110 where concrete in non-aggressive soil is defined as a moderate environment. Piles for highway structures should be designed in accordance with Table 13 of BS 5400: Part 4. In this standard, buried concrete is classified as being subject to severe exposure conditions or moderate for the case of concrete permanently saturated by water with a pH greater than 4.5. These conditions are more onerous than BS 8110. Nominal cover to reinforcement for Grade C30 concrete for buried parts of the structure is 45 mm for severe conditions or 35 mm for moderate conditions. However, BS 8004 recommends a minimum additional cover allowance of 40 mm when concrete is cast directly against an excavated soil face. The cement type, minimum cement content and maximum free water cement ratio shall be in accordance with BRE Digest 363 to protect buried concrete from acid and sulphate attack. Normally, the minimum clear spacing between bars for concrete placed by hopper or tremie should be 100 mm where aggregate of 20 mm maximum size is used. This spacing may be reduced to normally not less than 75 mm where aggregrate of 10 mm maximum is used.

4. BS 8004 limits the concrete stress in the shaft to 25% of the concrete characteristic strength. Where the casing of the pile is continuous and permanent, of adequate thickness and suitable shape, the allowable compressive stress in the shaft may be increased.

5. The length of pile required should be specified by the designer and may be amended following the results of preliminary pile tests. Trial bores may be useful prior to preliminary pile tests or installation of working piles to assess likely variations in soil conditions, bore stability and the permeability of soils to support fluid.

6. The requirement for the diameter of piles not to be less than the specified diameter is monitored by checking for concrete underbreak during concreting. Underbreak over a length of pile may indicate a defect. Apparent underbreak over a cased

length of pile with known dimensions is usually caused by the presence of an abnormally high percentage of steel and/or tubing in the pile shaft or an under-reporting of the volume of concrete supplied. Similarly, apparent overbreak may be caused by over-reporting the supply of concrete to the pile.

7. Typical bored cast-in-place pile diameters are given in Table 15 of BS 8004. Tripod bored piles generally are of 0.45 m or 0.6 m diameter and up to 25 m long (tripod bored piles longer than 20 m are subject to verticality problems in some ground conditions). Raking tripod piles in particular deviate quickly from their initial alignment and require a larger verticality tolerance. Auger bored piles commonly are bored at diameters ranging from 0.3 m to 2.4 m in 0.15 m increments but piles of much larger diameter (up to 3.6 m) are possible. Larger pile sizes (greater than about 1.5 m) will restrict the choice of Contractors.

8. The Contractor's proposals for concrete cast under support fluid should be checked to ensure that they meet the following:

   - Aggregates should preferably be naturally rounded well-graded gravels and sands when they are readily available in the locality. They must comply with BS 882. The maximum aggregate size should be 20 mm. The sand should conform with Grading M of BS 882. The use of other aggregates may be permitted subject to the suitability of the diameter of the tremie pipe and spacing of reinforcement bars.
   - A cementitious content of not less than 400 kg/m$^3$ should be maintained. Admixtures are often used to improve the workability, rate of gain of strength and setting time and these should comply with Section 20.
   - The mix should be designed to give high workability and the specified characteristic strength. The required strength depends on the loading and the requirements of BS 8110 to meet the anticipated exposure conditions. For reinforced concrete cast against soil and below water the minimum grade is C30 concrete.
   - The concrete cover to reinforcement should not be less than the values stated in Table 3.4 of BS 8110. For concrete cast against the ground an additional 40 mm of concrete cover is recommended by BS 8004 and thus the normal concrete cover to reinforcement is usually 75 mm.

9. Permanent casing may be used:

   (a) to provide support to zones of the ground surrounding the pile which may become unstable before the pile concrete is set;

   (b) to provide additional load carrying capacity to the pile in the situation where the concrete plus reinforcement is unable to provide the required load capacity, particularly the capacity to resist lateral loads;

   (c) as a surface on which slip coat may be spread (see Section 8);

   (d) as a barrier against the ingress of contaminated or aggressive groundwater. Such casing must be in firm contact with the soil surrounding the pile especially at the top and bottom of the section of shaft being cased. This can be done in small diameter shafts by expanding the casing, either by mechanical means or by fluid pressure, although this process is seldom used. In larger shafts the annulus should be filled with a 'thick' cement–bentonite grout;

(*e*) to form piles through water, e.g. in docks.

Generally, any significant annulus outside the casing should be filled with grout to prevent volume changes in the ground. The casing must be able to withstand the fluid pressure of the grout without buckling.

**10.** The more commonly encountered defects are summarized in CIRIA Report PG2. In particular, the requirement for a rigid 3 m long tremie pipe is to avoid the mix segregating if it hits the reinforcement cage.

**11.** Where the bases of piles are to be inspected by manned descent, safety must in no way be compromised. Some Contractors have views on manned pile descent which should be sought before such piles are specified. If a large diameter pile in a cohesive material is to derive a significant part of its resistance from its base, the base should be directly inspected as piling equipment smooths soft reworked material to make it appear to a closed circuit television system as if it is intact clay. Only piles in ground which is self-supporting and free from joints and seepages or where the piles are fully cased should be considered for descent. Although BS 8008 permits piles of diameter 750 mm to be descended for inspection purposes only, in most cases 900 mm piles will be the minimum diameter. The Contractor's method statement for manned descent should be carefully scrutinized to ensure the full requirements of BS 8008 are met.

**12.** Pressure grouting of the pile base or along a length of the shaft can be carried out in non-cohesive soils to enhance pile performance. The mechanism for pressure grouting of a pile base is discussed by Fleming (1993), who suggests that in addition to giving improved load–settlement characteristics, base grouting can also be used as quality assurance for bases which are required to carry load but cannot be directly inspected. The uplift limits specified for base grouting are typical for most pile types. For very long piles of small diameter, it may not be possible to measure any pile head movement at all. On the other hand, if it is possible to move a very long pile by 2 mm the movement at the pile toe may be considerably greater, causing the pile–soil interface strength to fall to a residual value in cohesive soils. In these circumstances the philosophy of base grouting needs careful consideration.

**13.** It is not possible to rake bored cast-in-place piles if the piles are to be concreted with a tremie pipe.

**14.** This Section can apply to mini- or micro-piles which are small diameter piles used for underpinning existing structures, or where working area is restricted, or for locations adjacent to sensitive structures, or where difficult ground conditions exist, such as boulders, fissured rock or man-made obstructions. The piles are usually of 100–300 mm diameter and up to 30 m long. Drilling equipment is normally employed and the pile filled with grout, sometimes under pressure to give additional penetration into granular soils. Reinforcement is usually a single bar held centrally in the drillhole by a special spacer.

15. Heavy mechanical breakers to cut down piles should be used with caution as they may induce damage below the point of application. This applies particularly to piles of less than 600 mm diameter.

# Section 4. Bored piles constructed using continuous flight augers and concrete or grout injection through hollow auger stems

1.  This section applies to bored piles which employ a continuous flight auger (cfa) for both advancing the bore and maintaining its stability. The spoil-laden auger is not removed from the ground until concrete or grout is pumped into the pile bore from the base of the hollow-stemmed auger to replace the excavated soil. The reinforcement is inserted after the pile has been concreted to the surface.

2.  The guidance in Section 3, guidance notes 2 to 6 and 15 for bored cast-in-place piles also applies to bored piles constructed using continuous flight augers with concrete or grout injection through hollow auger stems.

3.  The monitoring requirements for cfa piles are particularly onerous as cfa piles are the only pile type where the maintenance of pile bore stability is not observed. Where monitoring of key parameters has not been used, defects have occurred. It is necessary that the Contractor's monitoring system should be automated. In addition, it is desirable that the speed of rotation of the auger should be recorded as the number of auger rotations relative to auger penetration is a useful although not essential parameter. Measurement of the speed of auger rotation is not offered by many Contractors. The requirement for regular calibration provides confidence in the complicated automatic monitoring equipment that is often employed.

4.  The availability of a hard copy of the monitoring output after completion of a pile can provide a common basis for discussion should an incident occur during pile construction. This may allow the Contractor to evaluate rapidly the need for any remedial works before the concrete has set.

5.  Typical continuous flight auger piles range in diameter from 0.45 m to 0.9 m in 0.15 m increments and are generally up to 23 m deep below the commencing surface. Longer piles can be constructed by using two lengths of continuous flight auger, but difficulties can arise when splitting the auger (i.e. removing the upper length) during concreting when the concrete pressure in the pile bore reduces. This can have an effect on the integrity of the shaft and steps should be taken to ensure the Contractor guards against this. The specification does not permit auger joining during boring nor auger splitting during concreting.

6.  Piles constructed using a continuous flight auger have their reinforcement inserted on completion of concreting. The reinforcement is either pushed or vibrated into the concreted

*Specification for piling and embedded retaining walls.* Thomas Telford, London, 1996.

pile shaft. Difficulties may arise inserting heavy or long reinforcement cages or where there has been a delay between concreting and insertion. Particular problems are experienced in very permeable soils where the concrete stiffens more rapidly or where the reinforcement cage has a number of links or joints which resist its penetration.

7. During boring, when the auger passes from a weak stratum to a strong one, there is a danger that the weak soils will be drawn up the continuous flight by a process known as 'flighting'. This produces local shaft enlargement and possible loss of integrity. Flighting can also occur due to bulking of soils when excavated.

8. Permanent casing is not normally installed in conjunction with this pile type. This means there is no way of reducing friction along a length of pile.

9. With continuous flight auger piles, the soil is not seen until completion of construction. Therefore, it is difficult to measure penetration into a particular stratum or to observe whether or not a feature such as a dissolution feature in chalk has been encountered.

10. The requirement to rebore piles to 1 m below the original toe level if the auger is withdrawn from the ground during concreting is not practical if the original toe level was dictated by the presence of an impenetrable layer.

# Section 5.    Driven cast-in-place piles

1.  This section applies to piles for which a permanent casing of steel or concrete is driven with an end plate or plug, reinforcement placed within it if required and the casing filled with concrete. It also applies to piles in which a temporary casing is driven with an end plate or plug, reinforcement placed within it and the pile formed in the ground by filling the temporary casing with concrete before and sometimes during its extraction.

2.  The guidance in Section 2, guidance notes 5 to 9 for precast reinforced and prestressed concrete piles and precast reinforced concrete segmental piles, and Section 3, guidance notes 2 to 5 and 15 for bored cast-in-place piles, also applies to driven cast-in-place piles.

3.  Driven cast-in-place piles can be either top-driven (i.e. the hammer blows are applied to the pile head) or bottom-driven (where the blows are applied to a plug or steel plate at the base of the casing). Generally, Contractors will offer a proprietary system when this pile type is specified. A range of the systems available is summarized in CIRIA Report PG1.

# Section 6.   Steel bearing piles

1.  This clause covers steel piles driven to form part of the Works. Generally the piles will be hollow tubes, welded sections or H-sections. This section of the Specification and notes for guidance does not include steel sheet piling, except insofar as such piles are designed to act as bearing piles supporting vertical loads. Steel sheet piles are covered in Section 17.

2.  Considerable guidance on the specification and installation of steel bearing piles is given in Cornfield (1989) and British Steel (1992). In particular, this publication contains useful advice on design corrosion rates for different situations. However, the corrosion rates in BS 6349 should be designed for.

3.  Steel tubes can be driven open (i.e. without end plate) or closed (i.e. with end plate). If tubes are driven open, then the soil on the inside may 'plug', that is the soil inside the tube moves downward with the pile. If the piles are driven closed, an end plate is usually welded to the end. Such piles (or piles driven open with partial or total excavation of the soil inside) can be filled with reinforced concrete, in which case it is a driven cast-in-place pile with sacrificial lining or permanent lining. If the tube is driven open and the soil inside is subsequently excavated, considerable care is necessary not to induce inflows of water or soil into the tube.

4.  The guidance in Section 2, guidance notes 5 to 9 for precast reinforced and prestressed concrete piles also applies to steel piles.

5.  Clause 6.3 requires the Contractor to submit details of all preliminary test results to the Engineer at least 5 working days prior to ordering piles for the main work. Varying pile lengths, diameters or thickness of steel after the Contractor has placed his order could be expensive and cause delay. There is a significant risk of having to vary the order where piles are ordered before completion of the preliminary test piles, e.g. for rapid programming of the work.

# Section 7.    Timber piles

1.  This Section applies to new timber piles for inclusion as part of the permanent works.

2.  The guidance in Section 2, guidance notes 6 to 10 for precast reinforced and prestressed concrete piles and precast reinforced concrete segmental piles also applies to timber piles.

# Section 8.    Reduction of friction on piles

1. Where a means of reducing friction on any specified pile is required, one of the following methods can be used:

   (*a*)  pre-applied bituminous or other proprietary friction-reducing coating
   (*b*)  pre-applied low-friction sleeving
   (*c*)  formed-in-place low-friction surround
   (*d*)  pre-installed low-friction sleeving.

2. Pre-applied coating and sleeving are applicable to driven piles or permanently cased bored piles while the formed-in-place surround and pre-installed sleeving are applicable to driver or bored piles.

3. Most friction-reducing products are proprietary brands and care must be taken to follow the manufacturer's recommendations. In particular, for treatment of driven piles, the pile surface should be clean and dry before application of any coating.

4. The Contractor is required to provide the Engineer with a manufacturer's specification for any proprietary system used. The use of non-proprietary systems is not recommended as the requirements of a slip layer are both very demanding and partially conflicting, e.g. a pre-applied coating is required to remain attached to the pile during driving but is then required to shear during slow soil loading. In particular, the Engineer should ensure that the product used is compatible with the friction reduction assumed in the pile design.

5. Any set measurements or static or dynamic pile tests to prove friction reducing systems should allow for the different rate of load application applied in the test and those that will be applied by the soil to the working piles.

6. Care should be taken when driving piles with pre-applied coating or sleeving though coarse granular soils that the coating or sleeving is not removed.

# Section 9. Non-destructive methods for testing piles

**Integrity testing**

1. The purpose of integrity testing is to identify acoustic anomalies in piles that could have a structural significance with regard to the performance and durability of the pile. Integrity tests do not give direct information about the performance of piles under structural loads.

2. The methods available are normally applied to preformed concrete piles made in a single length, to steel piles and to cast-in-place concrete piles. The constituent materials of the piles should have a large differential modulus of elasticity compared with the ground in which it is embedded to obtain a satisfactory response. There is normally a limit to the length/diameter ratio of pile which can be successfully and fully investigated in this way depending on the ground conditions. Joints in piles, large changes in sections such as underreams and permanent casings are likely to affect the ability to obtain a clear response from the pile toe.

3. Integrity testing is not to be regarded as a replacement for static load testing but as a means of providing supplementary soundness information. Damage to the head of a pile after construction or cracks formed in the pile due to heaving of adjacent clay can often be detected. However, the structural significance of this cannot be reasonably assessed without subsequent physical examination of the pile to assess reductions in section, voids or crack widths, their location, orientation and the likely structural effects on the pile.

4. For sonic logging, it is preferable to install four tubes if the pile is of sufficient diameter. This allows the centre of the pile to be integrity checked. On completion of sonic logging it is normal for the tubes to be grouted up with a grout of comparable strength to the concrete in the pile.

5. Preparation of pile heads is required for most types of integrity testing, and as several tests can be carried out in a single visit, it is therefore necessary for the Contractor to take account of this in his programme.

6. Further guidance on integrity testing is provided elsewhere in this volume and in TRL Project Report 113 and CIRIA Funder's Report CP/28.

**Dynamic testing**

7. The purpose of dynamic pile testing is to determine the response of the pile to dynamic loading and to assess the efficiency of transfer of energy to the pile head from the hammer blow.

*Specification for piling and embedded retaining walls.* Thomas Telford, London, 1996.

8. Dynamic pile testing involves monitoring the response of a pile to a heavy impact applied at the pile head usually from the hammer used to drive the pile. The response is normally measured in terms of force and acceleration or displacement close to the pile head.

9. The results of the test directly obtained refer to the dynamic loading condition. The test is valuable for monitoring hammer efficiency, pile integrity for certain piles and driveability. It should not be considered as a complete substitute for static load testing of piles but can be a useful supplement once correlation between the static and dynamic tests has been established.

10. Tests are normally carried out on piles that have achieved their final set and the impact provided by the hammer is unlikely to be sufficient to move the pile far enough to mobilize the full pile resistance. Back analysis of dynamic tests to compare with static load tests should therefore be treated with extreme care.

11. If tests are required on restrike some time after installation to assess relaxation or set up effects this should be stated in the Particular Specification so that the Contractor can organize his programme accordingly.

12. The results required for typical blows are the output from a simple wave analysis program such as CASE. A more rigorous analysis may be required for selected blows. This will be accomplished using a program such as CAPWAP where the measured results are compared with a theoretical model built up from a knowledge of the soil properties and experience. The Engineer should seek to understand the output from both types of analysis as he may disagree with the input parameters leading to the stated results.

13. Further guidance on dynamic pile testing is contained elsewhere in this volume and in TRL Project Report 113.

# Section 10.    Static load testing of piles

1.  This section deals with the testing of a pile by the controlled application of an axial load. It covers vertical and raking piles tested in compression (i.e. subjected to loads or forces in the direction such as would cause the piles to penetrate further into the ground) and vertical or raking piles tested in tension (i.e. subjected to forces in a direction such as would cause the piles to be extracted from the ground). Static load testing of piles is the only reliable way to establish a pile's load–settlement behaviour.

2.  The purpose of preliminary piles is to validate the pile design and performance criteria and to prove that a Contractor's method of construction can construct viable foundations in particular ground conditions. Where preliminary piles are required they should be constructed sufficiently in advance of the installation of the working piles to allow time for the test, the evaluation of the results and the adoption of modifications if these prove necessary. If it is necessary to specify a precise timing for the construction and testing of preliminary piles this should be included in the Particular Specification. Normally, arrangements will be established when dealing with the Contractor's programme submitted under Clause 14 of the ICE Conditions of Contract, 6th Edition (or other conditions of contract as appropriate).

3.  A working pile may be tested at any time during the Contract but the potential benefit of choosing a pile from several that have been completed must be balanced against the risk of delaying the remainder of the Works if the pile proves to be unsatisfactory. Working pile tests are seldom tested to failure. Generally, working piles are tested to verify that the construction methods used have not changed so as to produce piles inferior to the preliminary piles, or to verify that piles which are for some reason suspect have satisfactory load–settlement performance. In some cases, such as with driven piles where relaxation or setting up may occur, it is necessary to delay the test until about 7 days after installation. For all cases it is preferable to test at least one working pile test to be confident that the requirements of the Specification have been met.

4.  Where the Contractor has designed the piles and performance criteria are specified, a test is an essential part of the Contract to establish that the piles meet the performance criteria.

5.  Preliminary piles should be specified unless the design, factor of safety against failure, construction method and ground conditions are such that the risk of failure of working piles is negligible. Further guidance on static load testing of piles is provided elsewhere in this volume.

6. Guidance on pile load testing procedures is given in CIRIA Report PG7. The methods of loading, i.e. under kentledge, against anchor piles or against ground anchors, should be submitted by the Contractor and not specified by the Engineer except in some instances for driven piles where the uplift of the test pile during driving of the anchor piles is of interest. Safety of the test arrangement is of paramount importance and the Contractor should give the Engineer the specified period of notice so that the test assembly can be inspected by the Engineer before the application of any load.

7. The Engineer should satisfy himself that the Contractor's method of load application will only impart an axial load into the pile, unless the load is on a laterally loaded pile.

8. The type of test should be stated in the Particular Specification as:

   - maintained load test as Clause 10.13.1
   - proof load test as Clause 10.13.1
   - extended proof load test as Clause 10.13.2
   - other systems such as cyclic loading or constant rate of loading test.

   Where other systems of test are stated, details should be provided of all the loading stages, measurement requirements and acceptance criteria.

9. The constant rate of penetration test option in Clause 10.13.2 is normally used where the ultimate load capacity of a preliminary pile is required, particularly for piles embedded in and bearing on clay soils. It may lead to apparently enhanced capacities by comparison with maintained load tests.

10. Special construction details may be required for preliminary piles in order to provide data that are relevant to working piles where downdrag is expected or which are piles which form part of a deep basement or substructure. Details may include sleeving and instrumentation within the pile. The method of construction should otherwise replicate as closely as possible the methods used to form working piles.

11. It is important that the Design Verification Load is appropriate to the situation of the test and the long-term loads for which the piles are being designed. As an example, if negative skin friction or downdrag is expected on the working piles, twice the expected downdrag force should be added to the Specified Working Load to give the Design Verification Load, once to overcome the positive skin friction over the relevant length and a second to replicate the actual downdrag loading. Where working piles are being installed in advance of an excavation, the Design Verification Load should take account of the support provided by the soil that will be excavated and also by the higher effective stresses giving higher strengths in the soils beneath.

# Section 11. General requirements for embedded retaining walls

1.  Sections 11 to 16 of this Specification are based on the draft Institution of Civil Engineers *Specification for cast-in-situ retaining walls*, with modifications. The requirements of this specification are very similar to the Highways Agency *Specification for highway works*, Series 1600, August 1994 issue.

2.  The document has been assembled so that the specifier selects the clauses of the Series considered to be appropriate for the Works and completes the appropriate Particular Specification for each clause selected. Section 11 will be required in all cases with other sections selected depending on the particular requirements of the structure, site, ground conditions and environmental constraints. A list of requirements to be specified is given in the Particular Specification. Other requirements can also be specified by adding sections to the Particular Specification.

3.  A comprehensive site investigation carried out in accordance with BS 5930 is an essential prerequisite to the design selection and construction of embedded retaining walls. Particular attention shall be given to the following aspects, which are particularly relevant to the execution of embedded retaining walls:

    *   a comprehensive desk study of the site history to assess the risk of obstructions, contamination, quarries, opencast and deep mines, backfilled sites or archaeological finds that could affect the feasibility of the wall construction, programme and cost. Suspected obstructions should be identified by probing or preliminary enabling works to confirm their extent;
    *   equilibrium piezometric levels of all possible water-tables, including artesian conditions and any seasonal or tidal fluctuations;
    *   permeability of the soils;
    *   presence of coarse, open soils, cavities, natural or artificial, which may cause sudden losses of support fluid in open excavation and require preliminary treatment;
    *   strength and deformation characteristics of soils, particularly weak strata that could cause instability or large deformation;
    *   the presence of boulders or obstructions which may cause difficulties in excavations or driving conditions;
    *   soil and groundwater chemistry which may affect the durability of the wall or the disposal of spoil and support fluid;
    *   the strength and profile of any rock surface along the wall alignment.

Conditions of Contract may affect what information should be provided in ground investigation reports and the responsibility attracted by the various parties particularly where interpretation is included in the reports. A check should be made that reports provided, which should include both factual and interpretative reports, are relevant to the structure to be built and that the scope of construction works envisaged at the time the site investigation was carried out have not significantly changed. If they have, further investigation should be carried out.

4. Options are provided for either the Engineer or the Contractor to design the embedded retaining wall. Guidance on responsibility for design is given elsewhere in this volume. Although the ICE document relates to piling, the general responsibilities and obligations apply equally to the design of embedded retaining walls. Similarly, the guidance given in Section 1, guidance note 3 also applies to embedded retaining walls. Reference to design should also be made to BS 8110 and BS 8004. The design of the wall should take into account the following:

- required design life;
- durability: crack control in reinforced concrete depending on the exposure conditions (see BS 8110, Table 3.4), the strength and permeability of the mix for soft secant piles, the corrosion of steel piles and the deterioration of timber lagging used for King Post walls;
- sequence of construction: the sequence of temporary or permanent propping and extent of stages of excavation;
- drainage: whether the penetration of the wall is to restrict groundwater flow and any provisions for drainage in front of the wall at formation level;
- watertightness: the designer and client need to have a clear understanding of what is required for the permanent works and what is reasonably achievable with the different forms of wall construction. BS 8102, Table 1 provides useful guidance on the grades of level of protection related to usage and the appropriate forms of construction. While BS 8102 is applicable to basement construction the basic principles can be considered for all excavations inside embedded retaining walls;
- structural loads and moments applied to the wall;
- soil and water pressures: these will vary according to the sequence of construction, temporary and permanent wall restraint and the effects of dewatering, over-excavation, erosion, etc. Reference should be made to BS 8002 and CIRIA Report 104;
- effects on surrounding structures: adjacent buildings, roads and utilities may be affected by ground movements associated with wall construction and the process of excavation in front of the wall. Vibration can cause settlements of loose granular soils, and dewatering may cause consolidation of cohesive soils and draw fines from cohesionless soils. Deflection of the wall during excavation of soil in front of the wall will cause settlement and horizontal movements of the retained soils.

5. Where the Engineer opts to design the wall only for the

permanent condition, the following information should be provided in the Particular Specification or on the Drawings:

- the assumed envelope of earth and water pressures acting on the wall;
- the propping forces provided by elements of the structure;
- the design life of the wall;
- the bending moment and shear force envelopes;
- typical reinforcement details or steel pile section moduli as appropriate;
- the assumed soil design parameters and water levels;
- required blockout and shear connection details;
- the layout, depth and thickness of the wall;
- required watertightness;
- limits on ground movements, vibration and noise.

The Engineer should ensure that the Contractor submits details of the wall elements in accordance with the relevant Particular Specifications as appropriate and details of any temporary works associated with the installation of the wall and subsequent excavation.

6. Where the Contractor is required to design the wall for the permanent condition, the Engineer should provide the following information:

- the design life of the wall;
- the location of the wall;
- required watertightness;
- design criteria; reference to BS 8110 and BS 8002 (or CIRIA Report 104 for walls embedded in stiff clays);
- limits on wall and ground movements;
- requirements for monitoring the wall (e.g. movement, noise and vibration);
- the permanent works which will incorporate the completed wall including any connections, facing details and the loads and moments that will be applied to the wall in the permanent condition from other parts of the structure;
- adjacent structures and utilities to be considered in the design including any limits on damage, noise and vibration, and requirements for monitoring and reporting.

The walling method statement should include sufficient detail to demonstrate that it is compatible with the design assumptions and any site constraints. The amount of detail required at the time of time of tender should be considered bearing in mind what is required for an initial assessment and comparison of tender proposals and what is reasonable for a Contractor to provide in a competitive situation. The Contractor's designs must be regarded as confidential (copyright laws apply) and parties receiving such designs should, if the Contractor so requests, be prepared to sign confidentiality agreements. The following list includes examples of items that may need to be provided:

- site staff and organization;
- the type of wall;
- the experience of the Contractor and his staff with the

walling type and the particular ground conditions;

- the type and number of rigs and ancillary equipment;
- the sequence of construction for the wall installation and subsequent excavation including all temporary and permanent conditions;
- wall dimensions comprising the sizes of panels, piles and lagging and including the spacing of these elements and corner details where appropriate;
- calculations for stability and performance of the wall to resist the stated pressures, load and moments to be applied in the temporary and permanent conditions of the wall, taking into account the stated sequence of construction;
- the material properties of the wall to meet the required design life and durability;
- the method of achieving the specified degree of watertightness and how this will be demonstrated on completion of the final excavation;
- the frequency and means of testing concrete workability and strength;
- details of reinforcement, steel sections and blockouts including details of additional reinforcement cover blocks and ancillary equipment required to maintain these items in their correct position during lifting, transportation and installation in the wall;
- the methods that will be used to check element depth and stability, base cleanliness if applicable and compliance with specified tolerances;
- equipment and method for monitoring surveys including frequency of readings and the form of presentation of results if not previously specified by the Engineer;
- trials in advance of the main wall installation to demonstrate that the method and materials will meet the Engineer's specified requirements. For example, these may include trial panels for diaphragm walls or trial mixes of concrete or slurry mixes for hard/hard or hard/soft secant piles;
- the method of achieving the specified verticality and positional tolerance including details of guide walls or guide frames with possibly radar or sonar being used to prove the excavated element profile;
- procedures for dealing with emergency situations such as sudden loss of support fluid, obstructions and panels or piles that are out of tolerance;
- company safety plan;
- quality assurance procedures;
- typical record sheets.

The Contractor must gain approval for his design (if applicable), proposed alternatives and/or construction methods from the relevant Authorities as necessary, e.g. a discharge consent from the Environment Agency if back of wall drainage is to enter a watercourse.

7. Unexpected and emergency situations can often arise with the construction of embedded walls as for piling. The guidance in Section 1, guidance note 5 applies equally to wall construction, and full-time supervision by the Engineer and Contractor is recommended.

8. The Engineer will normally determine the location and condition of adjacent structures and utilities that are likely to be within the zone of influence of the Works. Certain structures may be particularly susceptible to noise, vibration or ground movement, and the Engineer should assess prior to inviting tenders where special measures are likely to be required for protecting these structures. The information on location and condition of these structures together with the restrictions to be imposed and the monitoring requirements should be presented by the Engineer on the Drawings or in the Particular Specification. The Contractor is required to confirm the information on site and provide proposals to meet the requirements of the Engineer in respect of these structures and utilities.

9. The designer must ensure that the watertightness of the chosen walling method is achievable or that proper precautions for the secondary disposal of water ingress are incorporated, normally following the guidance of BS 8102 and CIRIA Report 139. Particular requirements should be specified in the relevant Particular Specification. In general, if seepage through the wall is of concern, secondary drainage measures should also be specified.

10. If it is not the Engineer, the body (Architect, Contract Administrator, etc.) acting as supervising officer should be specified in the Particular Specification. If that body's powers are not clearly defined elsewhere, then details should be provided there.

# Section 12.    Diaphragm walls

1. This section applies to diaphragm walls in which a trench (the excavation) is formed either by grabs or by reverse circulation cutters. Grabs using either rope operated or hydraulically operated clam shells advance the open excavation by removing material in separate bites while reverse circulation cutters allow almost continuous removal of material within the support fluid returns. Support fluid is used to support the walls of the trench prior to concreting. Each completed element is known as a panel.

2. Guide walls are used at the ground surface to ensure positional tolerance, to avoid surface erosion and to spread temporary loading. Guide walls are essential for maintaining the correct alignment of the wall and to provide a reservoir for the support fluid. The guide wall must be sufficiently robust to support any applied pressures from the ground and forces from the walling equipment. In particular, it should be able to support the surcharge from construction plant, the weight of the reinforcement cage if this is suspended off the guide wall and the reaction force from jacks if these are used to withdraw stop ends. The guide wall should be founded in soils of sufficient strength and stability to minimize the possibility of undercutting beneath the guide wall. The type and design of excavation equipment and the rate of withdrawal from the trench also influence the extent of potential undercutting.

3. The length of panels will depend on the Contractor's equipment, ground conditions, the proximity and loading from adjacent structures and vehicles or plant and permissible movements of the surrounding ground during construction. The stability of the panel relies on pressure from the support fluid within the trench exceeding the active earth pressure and water pressure within the soil arch adjacent to the panel. This surcharge of the ground from foundations and plant may require the hydrostatic pressure of the support fluid to be increased above the level that would otherwise normally be required (Clause 12.6.2) and the panel length restricted to reduce the earth pressure. Lengths of panels will normally vary between about 2 m and 7 m. Longer panels or special T or X shaped panels will normally require more than one tremie pipe to be used to provide an even spread of concrete across the wall section. Loss of support fluid leading to a reduction of fluid head in the excavation can have severe implications for the stability of panels. Careful consideration needs to be given to verticality and positional tolerances, particularly when considering steel design for reinforced T panels.

4. Where vertical loads are to be carried by the wall, the designer should take into account the practical level of cleanliness that can reasonably be achieved with the construction equipment as this could otherwise affect the performance of the wall. The

portion of wall above the final formation level is not normally considered as being able to resist vertical loads because of the reduction in horizontal stress and possible tension cracks that can occur in the soil behind the wall as the excavation is carried out. The time taken to construct panels will influence their vertical load capacity. If the panels are required to carry vertical load, the Contractor should be asked to provide the time within which he will have the panel excavated and concreted and this should be checked for compatibility with the design assumptions.

5. The Contractor's proposals for concrete cast under support fluid should be checked to ensure that they meet the following:

- Aggregates should preferably be naturally rounded well-graded gravels and sands when they are readily available in the locality. They must comply with BS 882. The maximum aggregate size should be 20 mm. The sand should conform with Grading M of BS 882. The use of other aggregates may be permitted subject to the suitability of the diameter of the tremie pipe and spacing of reinforcement bars;
- A cementitious content of not less than 400 kg/m$^3$ should be maintained. Admixtures are often used to improve the workability, rate of gain of strength and setting time and these should comply with Section 20;
- The mix should be designed to give high workability and the specified characteristic strength. The required strength depends on the loading and the requirements of BS 8110 to meet the anticipated exposure conditions. For reinforced concrete cast against soil and below water the minimum grade is C30 concrete;
- The concrete cover to reinforcement should not be less than the values stated in BS 8110, Table 3.4. For concrete cast against the ground an additional 40 mm of concrete cover is recommended by BS 8004, and thus the minimum concrete cover to reinforcement for diaphragm walls is usually 75 mm.

6. The joints between diaphragm wall panels are not normally watertight, particularly where there is a large difference in piezometric water pressure either side of the wall. Some proprietary joint systems include water bars to improve the degree of watertightness. The achievable degree of watertightness is best judged on the basis of the performance of similar wall systems in similar ground conditions. Where stringent criteria are specified the designer should consider whether additional measures such as grouting of joints or the provision of a facing wall with a drained and ventilated cavity as shown in BS 8102 should be provided to meet the requirements. Where a permissible inflow per metre squared has been specified, it should be measured over a 1 m by 1 m square area.

7. Preparation of wall surfaces will be required where drainage systems are to be installed and/or finishes applied. Details should generally be shown on Drawings and cross-referenced from the Particular Specification.

# Section 13.  Hard/hard secant pile walls

1. This section applies to hard/hard secant pile walls which consist of overlapping structural concrete piles constructed by high torque rotary piling equipment. Temporary support of the pile bore is provided by drill casing which generally extends over the full pile length unless continuous flight augers are used. The secant pile wall is constructed in two stages. All piles constructed during Stage 1 are known as primary (or female) piles. These are spaced at the specified primary secant pile spacing. All piles constructed during Stage 2 are known as secondary (or male) piles. These are positioned between the primary piles and secant (i.e. overlap) with the primary piles. Guide walls are usually used at the ground surface to ensure positional tolerance and initial pile bore stability.

2. The programme and sequence of construction of hard/hard secant piles is dependent on the requirement to form interlocking piles in a strictly controlled sequence of primary and secondary piles. The rate of gain of strength of primary piles affects the time within which secondary piles can be formed. High torque rotary drilling equipment provides more rapid construction than rigs where the casing is oscillated, and this rate of construction needs to be considered by the Contractor in conjunction with the rate of strength gain of primary piles to assess the sequence. Casing oscillators often have more power to cut through concrete.

3. The concrete mix may include additives to control the rate of gain of strength, particularly in primary piles. Where the Contractor considers that alternative proposals for the concrete mix are required then evidence of trial mixes or previous use should be provided. Where the specified characteristic strength is unlikely to be met at 28 days the Contractor should provide details of the anticipated rate of strength gain and the intervals between testing of concrete cubes and the time (say 56 days) when the specified strength will be met.

4. Particular care is needed with high powered rotary equipment to ensure verticality because of the rapid rate of construction. Oscillators often give better verticality. The Contractor should provide proposals prior to commencing the Works for remedial measures and revisions to the sequence of work in the event that a pile is formed outside the specified verticality tolerance. Oscillator casings have a thicker cutting edge to allow for cutting teeth.

5. The setting out and construction of guide walls require a high degree of accuracy because the cut crescent shape of the inside faces of the secant wall are critical for achieving the

correct centre to centre pile spacing and overlap. For a watertight wall, the primary pile spacing will be dictated by the need for pile overlap at the final excavation level; this will depend on the permitted pile position and verticality tolerances and depth of wall.

6. Where applied vertical loads or moments are to be supported by the wall, careful consideration should be given to the differential movement of adjacent piles and the effect on watertightness.

7. Secant pile walls have a greater frequency of joints than a diaphragm wall. Although these are cut concrete joints, in contrast to those of a diaphragm wall formed by a stop end, the watertightness of the secant wall may not necessarily be better than the diaphragm wall. The guidance in Section 12, guidance note 6 is also applicable to secant walls.

8. Steel sections are sometimes used in primary piles instead of reinforcing cages. These should be specified in the Particular Specification. In such cases where the concrete is placed by tremie, two pipes will be required: one on either side of the steel section.

9. Heavy mechanical breakers to cut down piles should be used with caution as they may induce damage below the point of application. This applies particularly to piles of less than 600 mm diameter.

10. The guidance in Section 12, guidance note 7 also applies.

*Specification for piling and embedded retaining walls*. Thomas Telford, London, 1996.

# Section 14.    Hard/soft secant pile walls

1. This section applies to hard/soft secant pile walls which consist of overlapping concrete piles constructed by rotary piling equipment. Temporary support of the pile bore may be provided by drill casing which may extend over the full pile length. Alternatively, a continuous flight auger technique may be used for some or all of the piles. The secant pile wall is constructed in two stages. All piles constructed during Stage 1 are known as primary (or female) piles. These are spaced at the specified primary secant pile spacing. All piles constructed during Stage 2 are known as secondary (or male) piles. The primary piles are formed of a low strength bentonite/cement mix which can be easily cut away by the secondary piles which are formed of reinforced concrete. The secondary piles are positioned between the primary piles and secant (i.e. overlap) with the primary piles. Guide walls are usually used at the ground surface to ensure positional tolerance and initial pile bore stability.

2. The guidance in Section 13, guidance notes 5 to 10 also apply to hard/soft secant pile walls. The decreased verticality tolerances and reduced watertightness of a hard/ soft secant wall compared with a hard/hard secant wall must be considered.

3. The durability, performance and design life of hard/soft secant pile walls require particular consideration in relation to the design mix of the soft piles. The soft piles are normally formed with a weak bentonite/cement mix with an undrained shear strength of the order of 0.2 N/mm$^2$ or 0.3 N/mm$^2$ (200–300 kN/m$^2$) which acts as a void filler. It therefore does not comply with normal durability requirements. It is likely that the properties of the bentonite/cement mix will vary more widely than conventional cement-based mixes and allowance for this should be made in the specified requirements for the mix.

# Section 15. Contiguous bored pile walls

1.  This section applies to contiguous bored pile walls which consist of bored piles constructed by rotary piling equipment or continuous flight augers. The piles are constructed at centres equal to the pile diameter plus an allowance for temporary casing width and tolerance which can vary between 70 mm to 150 mm, although in soils which are capable of standing vertically a larger spacing may be adopted. Temporary support of the pile bore is provided by casings and if necessary support fluid at lower levels. Guide walls may be used at ground level to ensure positional tolerance.

2.  Where stiff clay is to be retained, the designer may consider increasing the pile spacing and provide an in-situ reinforced concrete or blockwork facing to be applied during or on completion of excavation. Provision for vertical drains behind the facing between the piles should be considered.

3.  Guide walls are recommended where more stringent verticality and positional tolerances equivalent to diaphragm or secant pile walls are required.

4.  Contiguous bored piles do not retain water. If the wall is required to retain water then either the space between the piles can be grouted before excavation takes place, or a facing wall added (which may have a drained and ventilated cavity behind, as shown in BS 8102). It should be remembered that grouting between contiguous piles is unlikely to provide as watertight a seal as a secant pile or diaphragm wall.

5.  The guidance in Section 13, guidance notes 9 and 10 also applies.

*Specification for piling and embedded retaining walls*. Thomas Telford, London, 1996.

# Section 16. King Post walls

1. This section applies to King Post walls, also known as 'Berlin Walls' or soldier pile walls, most often used as temporary upholding works and which may be anchored depending on site conditions. The wall consists of drilling bores of a specified diameter at specified centres, usually between 2.5 m and 4.0 m. Posts, steel beams or pre-cast concrete sections are placed in the bores, and their lower section, up to excavation level, are concreted into the bore. Temporary casings extending to just above excavation level support the pile bore and the posts until the concrete has set, normally approximately 12 hours. The bore above concrete level is backfilled with spoil or in some cases grout, and the temporary casing is extracted. During excavation on one side of the wall, poling boards or some other form of lagging are placed between the King Posts to retain the soil.

2. Where permanent walls are to be constructed in front of a temporary King Post wall and the space between is to be filled with compacted material, consideration should be given to the sequence of removal of props or destressing of anchors, the durability of the King Post wall materials and the effect of compaction on the permanent wall. The materials forming the King Post wall may be removed where practical, provided that the method is such that damage is not caused to the permanent works, adjacent structures, highways and utilities.

3. Relaxation and some ground loss is inevitable in most cases during the excavation and installation of the lagging between the King Posts. The wall is permeable and hence groundwater levels behind the wall will be drawn down. These effects can cause damage to structures, highways and utilities if they are near the wall. If damage is likely and unacceptable the Engineer should not permit this form of construction, and should consider the diaphragm, secant pile or sheet pile wall alternatives.

4. The concrete is normally unreinforced. The workability should be sufficient to allow the concrete to flow around the King Post and adequately to fill the bore. Where concrete is placed under water or support fluid, it should have high workability. Two tremies may be needed to ensure that concrete flow is uniform either side of the King Post.

# Section 17.    Steel sheet piles

1. This section covers interlocking steel sheet pile sections which are generally constructed to resist horizontal soil and water pressures and/or imposed loads. Sheet piles are used in both temporary and permanent situations for cofferdams, retaining walls, bridge abutments and dock and harbour works. They are usually installed by use of impact hammers, vibrators or hydraulic jacking methods. Steel sheet piles have interlocking clutches which are capable of sliding together to form a continuous structural retaining wall. They are produced by hot rolling steel blooms in a rolling mill to form the finished piling sections.

2. Guidance on corrosion and protection of steel piling is given in BS 8002. Guidance on the effective life of unpainted or otherwise unprotected steel piles is also given in the *British Steel piling handbook*. Measures such as coatings, cathodic protection and increased steel grade or thickness to increase the design life in potentially corrosive environments are also described. Further information on design rates of corrosion are given in Cornfield (1989), British Steel (1992) or BS 6349.

3. The rolling dimensional tolerance of the sheet piles, particularly that of the pile clutch, can be variable from manufacturer to manufacturer. When long sheet piles are to be driven, it may become necessary to consider specifying the interaction factor of the clutch in the design in order to reduce the incident of declutching.

4. Sections of 'Z' type steel sheet piling (e.g. Frodingham) which have their interlocks in the flanges develop the full section modulus of an undivided wall of piling under most conditions. Sections of trough or 'U' type steel sheet piling (e.g. Larssen) with close-fitting interlocks along the centre-line or neutral axis of the sheeting develop the strength of the combined section only when the piling is fully driven into the ground. The shear forces in the interlocks may be considered as resisted by friction due to the pressure at the walings and the restraint exercised by the ground. In certain conditions it is advisable to connect together the inner and outer piles in each pair by welding, pressing or other means, to ensure that the interlock common to the pair can develop the necessary shear resistance. Such conditions arise when:

   (*a*)  the piling passes through very soft clay or water;
   (*b*)  the piling is prevented by rock from penetrating to the normal depth of cut-off;
   (*c*)  the piling is used as a cantilever; or
   (*d*)  the piling is supported by props or struts but is cantilevered to a substantial distance above the highest waling or below the lowest waling.

If any of these conditions arise and the piles are not connected together into pairs as described, a reduced value of section modulus of the combined section should be used.

5.  Sheet pile sections should be selected to ensure that they are capable of withstanding the bending moments and axial loads which will be applied during and after construction and that they are capable of being driven successfully through the soil to the required penetration. Guidance on pile selection is given in the *British Steel piling handbook*. Jetting can be used to assist pile driving in cohesionless soils but this may affect the performance of the piles under the applied moments and loads. Jetting can also wash away soil supporting surrounding structures.

6.  Where noise and vibration are of concern, jacking systems are available for installing sheet piles. Some proprietary systems can be used only in ground conditions predominantly comprising stiff clays; other systems can push piles through cohesionless soils as well but may be limited in the penetrations they can achieve.

7.  For handling purposes, sheet piles of the U type are normally supplied and pitched singly, while piles of Z type are normally supplied and pitched in pairs.

# Section 18.    Integrity testing of wall elements

1. The purpose of integrity testing is to identify acoustic anomalies in wall elements that could have a structural significance with regard to the performance and durability of the wall element. Integrity tests do not give direct information about the performance of wall elements under structural loads or their watertightness.

2. The sonic logging method is the only meaningful method for concrete wall elements other than contiguous pile walls — results of other techniques will be difficult to interpret and may be misleading. For contiguous pile walls, the techniques given in Section 9 can be used and the advice in that section's guidance notes followed.

3. For sonic logging, it is preferable to install four tubes in piles if they are of sufficient diameter. This allows the centre of the pile to be integrity checked. On completion of sonic logging it is normal for the tubes to be grouted up with a grout of comparable strength to the concrete in the pile.

*Specification for piling and embedded retaining walls.* Thomas Telford, London, 1996.

# Section 19. Instrumentation for piles and embedded walls

1.  This section applies to instrumentation for piles and for embedded walls. The instrumentation can be used for monitoring stresses or displacements in piles either in preliminary load tests or in service and for lateral and vertical loads. The results can be used to derive the load and deflection distribution down the pile or wall.

2.  Care should be taken before welding strain gauges to steel to ensure that the type of steel can be welded. Care should also be taken that the welding heat does not cause damage to waterproof seals at either end of the strain gauge.

3.  It is very important that one person is responsible for all aspects of instrumentation and has the time to coordinate all the necessary activities. Otherwise, if sufficient care is not taken throughout the installation, reading and interpretation processes, the results can become impossible to interpret.

4.  Thought should be given when specifying instrumentation as to how exactly the results will be used. As an example, if inclinometers in a retaining wall are being monitored, it will be necessary to know also the depth and extent of the excavation at each set of readings.

*Specification for piling and embedded retaining walls.* Thomas Telford, London, 1996.

201

# Section 20.   General requirements for reinforced concrete

1. If the specific gravity of either the coarse or the fine aggregate differs significantly from 2.6, the weight of each type of aggregate shall be adjusted in proportion to the specific gravity of the materials.

2. The cement shall be ordinary Portland cement or sulphate-resisting Portland cement. Where other cements in Clause 20.2.1 are used, special consideration shall be given to the design of the mix.

3. The weights of cement and dry aggregates are those which will produce approximately one cubic metre of compacted concrete.

4. Fine aggregates which consist mainly of angular particles are to be avoided.

5. The sand aggregate gradings given above are as specified in BS 882. The mixes described may require some adjustment on inspection where the gradings, although within the specified limits, are near the limits of the grading zone. Once a satisfactory prescribed mix has been adopted, the sources of aggregate shall not be changed.

6. The concrete produced should be cohesive and in a slump test should not shear, fragment or segregate easily.

7. BS 1881, Part 102: *Method for determination of slump* states that the slump test is not suitable for determining workability where the measured slump exceeds 175 mm. For very high workability concrete, such as conditions requiring piling mix workability C in Table 3.1, the use of the flow table (to BS 1881 Part 105) is recommended. However, much experience has been built up using the slump for very high workability mixes and Contractors often have trouble persuading their concrete suppliers to provide concrete whose flow rather than slump is controlled. More experience needs to be built up. The method of workability measurement can be specified in the relevant Particular Specification.

# Section 21. Support fluids

1. Where out of balance water forces may lead to instability of essentially dense or stable materials, water may be used to maintain stability provided that it does not adversely affect any design criteria such as soil strength. If water is used as a support fluid then the other requirements for support fluids do not apply.

2. The stability of the pile bore or wall trench during excavation can be maintained using support fluids where the fluid pressure acts against the sides of the excavation to support the surrounding soil and water pressures. In the case of bentonite slurries, this action is achieved by the formation of a filter cake in contact with the soil against which the fluid pressure acts. In the case of polymers, this action is due to rheological blocking of the fluid in the soil; the penetration distance is small in the case of clayey soils but can be significant in the case of sandy or silty soils. The main factors affecting the stability which can be controlled during the excavation are:

   - the level and properties of the supporting fluid
   - the size of the panel for diaphragm wall excavations
   - the time during which the pile bore or trench is left open relative to the soil and groundwater conditions (loss of shear strength with time)
   - the care that is taken not to create suctions with the excavation plant in the support fluid filled excavation.

3. Bentonite is commonly used as a support fluid and in its natural form as sodium montmorillonite exhibits thixotropic properties, whereby it forms a gel under quiescent conditions and regains its fluidity under dynamic conditions. Naturally occurring sodium montmorillonite is relatively rare and it is therefore more common to use activated bentonite which is a calcium montmorillonite that has been converted to provide similar thixotropic properties to sodium bentonite. Additives in the form of dispersants and deflocculants are sometimes used to improve the flow characteristics of the fluid, and viscocifiers or flocculating agents are used to improve the gelling or blocking action of the fluid. Further guidance on bentonite and its use and behaviour in the construction of bored piles is described in CIRIA Report PG 3.

4. Support fluids can also be based on other substances such as polymers.

5. Hutchinson *et al.* (1975) report on the various tests available to measure the properties of bentonite and list appropriate tests. These consist principally of the density, plastic viscosity, shear strength, filtration or fluid loss properties, pH and sand content. Knowledge of these properties is relevant to the following requirements of the fluid:

- solid particles are kept in suspension
- the fluid can be easily displaced during concreting
- continuous support of the excavations, for fluids which provide support by rheological blocking.

The concentration of bentonite needed to achieve the above requirements depends on the source of bentonite as described by Hutchinson *et al.* (1975), ranging from 3% for natural sodium montmorillonite to 7% or more for some manufactured bentonites. For activated calcium montmorillonite manufactured in the UK, a concentration of 5% has generally been found to be appropriate.

6. Table GN21 is provided as a guide to the type of test and compliance values that would normally be expected to be provided by the Contractor for activated bentonite manufactured in the UK.

The tests are described in the American Petroleum Institute document: *Recommended practice standard procedure for field testing water-based drilling fluids.* Compliance should be checked for the support fluid as supplied to the excavation and prior to concreting because the test requirements of the supporting fluid are different when supporting the excavation compared with the flow conditions needed for effective concreting. Further guidance on these aspects is given in CIRIA Report PG3 which also identifies the following practical limitations to the construction of excavations under bentonite suspension:

- very permeable strata causing loss of bentonite suspension which prevent the maintenance of the correct suspension level (soil of permeability up to $k = 10^{-3}$ m/sec can be stabilized with bentonite suspensions having concentrations of up to 6% by weight);
- cavities which may lead to sudden or excessive loss of bentonite suspension;
- very weak strata, such as some estuarine clays, with cohesion values of less than 10 $kN/m^2$ (very weak strata present a problem with retention of fresh concrete and a casing or liner may be required even if satisfactory conditions during excavation can be obtained);
- water under artesian head.

*Table GN21. Tests and compliance values for support fluid prepared from bentonite manufactured in the UK*

| Property to be measured | Test method and apparatus | API RP13 Section | Compliance values measured at 20°C | |
|---|---|---|---|---|
| | | | As supplied to pile | Sample from pile prior to placing concrete |
| Density | Mud balance | 1 | Less than 1.10 g/ml | Less than 1.15 g/ml |
| Fluid loss (30 minute test) | Low temperature test fluid loss | 3 | Less than 40 ml | Less than 60 ml |
| Viscosity | Marsh cone | 2 | 30 to 70 seconds | Less than 90 seconds |
| Shear strength (10 min gel strength) | Fann viscometer | 2 | 4 to 40 $N/m^2$ | 4 to 40 $N/m^2$ |
| Sand content | Sand screen set | 4 | Less than 2% | Less than 2% |
| pH | Electrical pH meter to BS 3445; range pH 7 to 14 | - | 9.5 to 10.8 | 9.5 to 11.7 |

7. Contamination of the bentonite can be detected by the tests described above and these need to be considered with regard to the effects on construction and performance of the pile or wall. Fine sand in the slurry during excavation may assist the blocking mechanism. However, the increase in density and viscosity due to the presence of sand may affect the flow properties and hence the ability of the concrete to displace the fluid during concreting. Also, any sediment forming on the base of the excavation may affect the performance of the pile or wall under load if the end bearing component is significant. It is recommended that the sand content should be limited to 2% prior to concreting if working loads are to be partly resisted by end bearing. Contamination with clay, usually in the calcium or aluminium forms of bentonite, can promote ion exchange with the slurry to such an extent that the filter properties are markedly changed. Cement contamination has a similar effect and can be detected by an increase in pH. The shear strength of the slurry is important in keeping small particles in suspension and hence avoid the formation of sediment at the base of the excavation.

8. The replacement of helical reinforcement by hoops or bands of steel spaced well apart is recommended to minimize accumulation of bentonite residue on the intersection of longitudinal and shear reinforcement.

9. The testing of the support fluid should be carried out in a properly equipped site laboratory. Samples of support fluid for testing should be taken from the base of the excavation. Where the test results indicate non-compliance with the stated limits the fluid should be partly or entirely replaced before concreting or further excavation takes place.

10. Current legislation affecting the disposal of support fluid includes the Water Act 1989 and the Environmental Protection Act 1990.

# References

American Petroleum Institute. *Recommended practice standard procedure for field testing water-based drilling fluids.* API Recommended Practice 13 B-1, 1990, June.

Building Research Establishment. *Sulphate and acid resistance of concrete in the ground.* BRE, 1991, July, BRE Digest 363 .

British Standards Institution. British Standard 882: *Specification for aggregates from natural sources for concrete.* BSI, London.

British Standards Institution. British Standard 5228: *Noise control on construction and open sites, Part 4: Code of practice for noise and vibration control applicable to piling operations.* BSI, London.

British Standards Institution. British Standard 5400: *Steel, concrete and composite bridges.* BSI, London.

British Standards Institution. British Standard 5930: *Code of practice for site investigations.* BSI, London.

British Standards Institution. British Standard 6349: *Code of practice for maritime structures.* BSI, London.

British Standards Institution. British Standard 8002: *Code of practice for earth retaining structures.* BSI, London.

British Standards Institution. British Standard 8004: *Code of practice for foundations.* BSI, London.

British Standards Institution. British Standard 8008: *Guide to safety precautions and procedures for the construction and descent of machine-bored shafts for piling and other purposes.* BSI, London.

British Standards Institution. British Standard 8102: *Code of Practice for protection of structures against water from the ground.* BSI, London.

British Standards Institution. British Standard 8110: *Code of practice for structural use of concrete.* BSI, London.

British Steel. *British Steel piling handbook.* British Steel, 1988, 6th edn.

British Steel. *The corrosion and protection of steel piling in temperate climates.* British Steel General Steels, 1992, Oct., publication P115.

Weltman A.J. and Little J.A. *A review of bearing pile types.* CIRIA, Report PG1, Jan. 1977.

Thorburn S. and Thorburn J.Q. *Review of problems associated with construction of cast-in-place concrete piles.* CIRIA, Report PG2, Jan. 1977.

Fleming W.G.K. and Sliwinski Z.J. *The use and influence of bentonite in bored pile construction.* CIRIA, Report PG3, Sept. 1977.

Weltman A.J. *Pile load testing procedures.* CIRIA, Report PG7, Mar. 1980.

Padfield C.J. and Mair R.J. *Design of retaining walls embedded in stiff clay.* CIRIA, Report R104, 1984.

Construction Industry Research and Information Association. *Water-resisting basements.* CIRIA, Report 139, 1995.

Construction Industry Research and Information Association. *The role of integrity and other non-destructive testing in the evaluation of piled foundations.* CIRIA, Funder's Report CP/28, 1995.

Cornfield G.M. *Steel bearing piles.* Steel Construction Institute Publication 016, 1989, 4th edn.

Fleming W.G.K. The improvement of pile performance by base grouting. *Proc. Instn Civ. Engrs Civ. Engng*, 1996, May, 88–93.

Hammond A.J., Mitchell J.M. and Lord J.A. Design and construction of driven cast in situ piles in stiff fissured clays. *Proc. Conf. on Recent Developments in the Design and Construction of Piles*. Thomas Telford, London, 1980, 157–168.

Hutchinson M.T., Daw G.P., Shotton P.G. and James A.N. The properties of bentonite slurries used in diaphragm wall and their control. *Proc. Conf. on Diaphragm Walls and Anchorages*. ICE, London, 1975, 33–40.

Institution of Civil Engineers. *Specification for piling*. Thomas Telford, London, 1988.

Tomlinson M.J. *Foundation design and construction*. Longman Scientific and Technical, 1995, 6th edn.

Transport Research Laboratory. *Advice on integrity testing of piles*. TRL, Project Report 113, 1995 .